METAL CASTING: A SAND CASTING MANUAL FOR THE SMALL FOUNDRY VOL. I

STEPHEN D. CHASTAIN

B.SC. MECHANICAL ENGINEERING AND MATERIALS SCIENCE
UNIVERSITY OF CENTRAL FLORIDA

Metal Casting: A Sand Casting Manual for the Small Foundry Vol. I
By Stephen D. Chastain

Copyright© 2004 By Stephen D. Chastain

Jacksonville, FL All Rights Reserved

Printed in USA, First Edition 3rd Printing

ISBN 0-9702203-2-4

ISBN 13: 978-0-9702203-2-5

Additional Titles by Stephen Chastain:

Metal Casting Volume 1

Metal Casting Volume 2

Iron Melting Cupola Furnaces

Small Foundry Furnaces Volume 1: Build A Tilting Furnace

Small Foundry Furnaces Volume 2: Build A Rotating Furnace

Sand Machinery Volume 1: Build a Muller

Making Pistons For Experimental and Restoration Engines

Generators and Inverters: Building Small Combined Heat and Power Plants

Web site: StephenChastain.Com

WARNING – DISCLAMER

This book is to provide information on the methods the author used to operate a small foundry. Foundry work can be dangerous. No attempt has been made to point out all of the dangers or even a majority of them. Although the information has been researched and believed to be accurate, no liability is assumed for the use of the information contained in this book. If you do not wish to be bound by the above, you may return the book for a full refund.

Warning: Molten metal and high intensity combustion can be dangerous. Incomplete combustion produces carbon monoxide, a poisonous gas. Only operate a furnace outdoors. Stay clear of all ports when a furnace is in operation. Observe all rules regarding safe foundry practice. Do not attempt to melt metal if you are not qualified. Do not use gasoline or other low flashpoint fuels to light a furnace. Do not spill molten metal on yourself, others or any wet or damp surface. Always wear protective gear. Observe all regulations regarding the safe handling of gaseous and liquid fuels. Safety is your primary responsibility.

Table of Contents Volume 1

Purpose & Introduction	5
Health and Safety	7
I. THE SAND CASTING PROCESS	9
The sand casting process	9
Basic Foundry Equipment	10
II. FOUNDRY PROJECTS:	21
Make a Match Plate Vibrator	21
Mold Making	29
Making Wooden Flasks	40
Making Flask Hardware	47
Making an Aluminum Flask	54
Rammer	57
Hardware Patents	59
III. MELTING EQUIPMENT	61
Crucibles and Ladles	61
Furnaces for the Small Foundry	70
Beginner's Charcoal Furnace	75
Stepping up to a Gas Fired Furnace	84
Forming Sheet Metal for Furnaces	98
IV. TEMPERATURE MEASUREMENT	113
Temperature Measurement	113
Thermocouples	114

Making a Thermocouple Thermometer	117
Optical Pyrometers	121
V. FOUNDRY SANDS AND BINDERS	126
Molding Sand	126
Clays	145
Mullers and Mulling	148
Cement Binders	150
Development of Oil Bonded Sand	153
Petro Bond	154
Furan Binders	158
VI. COREMAKING	160
Cores	160
Core Binders	169
Baking Cores	173
Core Finishing	179
Core Jigs	183
Core Buoyancy	185
Strainer Cores	191
Core Coatings	193
BIBLIOGRAPHY	196
APPENDIX	197
SUPPLIERS	202
INDEX	204

Continued In Volume II

PURPOSE:

The purpose of this book, volumes I and II, is to describe and demonstrate the basic sand casting process so it may be successfully applied to the small foundry. The book will cover green sand molding, skin dried and dry sand molds. Sands, core practice, mold washes and basic molding equipment are described, as well as the basic principles of gating and risering. A brief introduction to the metallurgy of cast iron, copper based alloys and aluminum are presented. Construction of small furnaces and various pieces of small foundry equipment are included so that you can build your own equipment as your foundry grows. Temperature measurement, an introduction to pattern making and typical automotive casting of the 1930's complete the sand casting books.

This book is not intended to be an exhaustive study of foundry practice. It is deliberately focused on low-tech readily available binders such as clay, linseed oil and molasses. Man has cast metals for thousands of years before modern polymer binders, petro-bond, cast Styrofoam patterns, and air set sand were available. Your grandfather probably drove a car with an engine cast in a skin dried mold sprayed with molasses water and pasted up cores made of sand and linseed oil. They work as well today as they did then.

Metal casting has been an art long before it was a science. In some areas the science has yet to catch up with the art. Basic theories are presented to help you design or troubleshoot your work as your casting projects become more complicated. As with all the books in the "Small Foundry Series," math is included as a tool or to help explain an idea. The math is not difficult and may be done on $9.95 scientific calculator (read the instructions for the calculator). Tables and graphs are provided where calculations are long and tedious.

INTRODUCTION:

My interest in metal casting goes back as far as I can remember. As a boy we lived in a small house located in central Florida. My father, a struggling young dentist, had a card table set up in my bedroom where, every evening, he would carve the wax for investment casting of crowns and bridges. I was fascinated by the process and later worked in his lab. I would cast several models that he would inspect. Some he would accept and some he would reject. They all looked perfect to me and I could never figure out *"why"* some were good and some were not. One day he came in and said "Steve, you've got to quit putting the bottom teeth on the top and the top teeth on the bottom!" I finally knew *"why."* Problems are easy to correct or avoid completely if you know why things are done a certain way.

Most casting books are either extremely basic or full of partial differential equations, neither of which is much help explaining "why" to the small foundryman. This book describes the process to get you started and explains the basic theory behind it.

Reading a book about metal casting does not make you an expert metal caster any more than reading a book about piano playing make you a virtuoso pianist. It takes practice. You will not learn metal casting in a few days. All good things require effort. When you run into a problem refer back to this book to find out "why." You will soon be able to produce good castings repeatedly and often on the first attempt.

HEALTH AND SAFTEY ISSUES:

There are obvious and not so obvious safety issues regarding foundry work. Obviously you can be burned by spilled or spattered metal; however silicosis, is a less obvious problem.

Silicosis is the hardening, scarring or stiffening of the lungs due to irritation caused by the inhalation of silica dust. Almost any type of dust may become a hazard when airborne. Dust in the non-visible range is the most hazardous. Larger dust particles become lodged in the nose, throat, and larger passages of the lungs where they are picked up by mucous and removed. Some of the smaller particles become trapped in the air sacs or alveoli of the lungs causing the lung tissue to form nodules. Long term exposure to dusty conditions on an almost daily basis for extended periods of time may result in silicosis. Symptoms including coughing wheezing and chest pains that may take 15 or 20 years to appear. Chronic silicosis may take 20 to 45 years to appear.

Prevention of silicosis depends entirely on prevention of inhalation. Proper ventilation and dust control systems are essential. There is no treatment for silicosis. NIOSH approved masks, vacuuming work clothes before removal and showering to remove dust from your skin reduces your exposure. Washing your hands before eating and drinking will further reduce your risk. Smoking greatly increases your risk.

In foundry work, the dustiest operations include mixing of dry sands, shakeout of castings, sand screening and grinding. Dust can sometimes be suppressed by wetting the dust sources or by using sprays to remove the airborne dust at the source. Silicosis in younger people is often linked to extreme overexposure during sandblasting.

Of the approximately 300 silicosis deaths per year, 76% are in people 65 and over, 23% are in people 45 to 64 and

1% in 15 to 44 years of age. For further information regarding silicosis contact OSHA.

Hot metal can be hazardous if spilled or if it comes in contact with even a small amount of water. Never allow wet or damp material to come in contact with molten metal. Steam will be violently liberated possibly causing an explosion, spattering molten metal around the work area. Do not allow molten metal to run over concrete. Always work over a bed of dry sand. Always wear protective gear. Long sleeves, welder's gloves, goggles and a facemask are important. A cap for your head and leather boots without laces are a good choice. If you wear shoes with laces then you should wear sheetmetal leggings and covers or spats over your shoes. In case of an accidental spill on your shoe they prevent molten metal from running down into your shoe and burning your foot.

While zinc oxide is common in many skin creams and considered nontoxic, when zinc oxide is inhaled it causes a condition known as "brass shakes." The condition was common at old zinc smelters and during the melting of high zinc alloys. The symptoms include a dry throat, hacking cough, dull headache, and nausea. A slightly chilly feeling occurs causing the person to shake. Muscular pains accompany the chill. After a few hours, the chills cease rather suddenly and the attack is over. A typical remedy was to drink large quantities of hot milk.

Prevention of the brass shakes depends upon prevention of inhalation of zinc oxide. Modern foundries pour under a fume hood that vacuums away the zinc fumes. Covered ladles also reduce the amount of zinc oxide in the air. Respirators may be used, however many feel that they are uncomfortable.

Keep combustible materials a safe distance away from the foundry operation. Always be on the lookout for hazards.

THE SAND CASTING PROCESS:

Green sand casting is not difficult and it is not too expensive to get started. Molten metal is taken from a furnace and poured into a sand mold. The sand in the mold is held together by a binder such as clay and a little water to make a damp sand or **green sand** mold. The mold is made by packing molding sand around a **pattern** or duplicate of the part to be cast. A **flask** holds the sand around the pattern. Once the sand is packed or **"rammed"** around the pattern, the mold is opened and the pattern is removed. If required, sand **cores** are set into place. The mold is closed and the molten metal is poured into a **basin** or **pouring cup**. It then flows through a hole made in the sand called a **sprue**. The molten metal flows into the mold cavity through a **gate**. After the metal has cooled, the **casting** is removed from the sand, cleaned and finished.

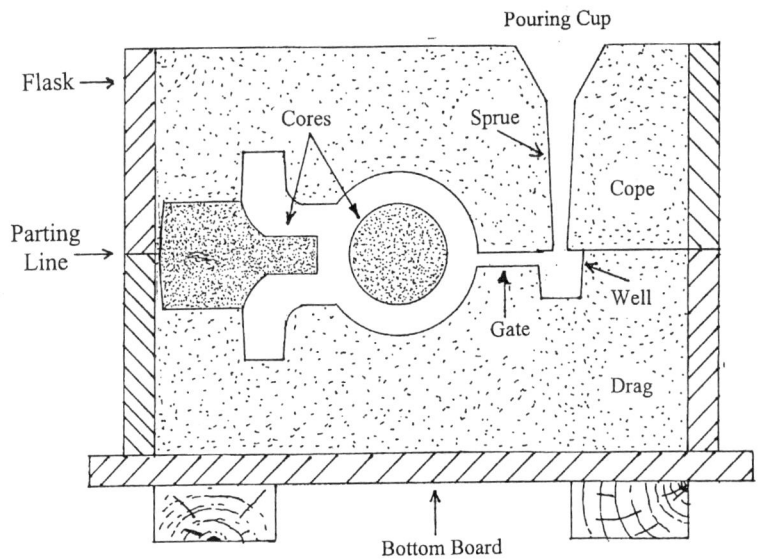

Sand Mold

BASIC FOUNDRY EQUIPMENT:

Flask: A flask is a frame or box that keeps the sand in place while the mold is being made. The flask has a top section called a cope and bottom section called a drag. If there are sections between the cope and drag they are called cheeks. Cheeks may be used to mold tall parts. Flasks may be made of wood or metal. The flask may be left on the mold while it is being poured or it may be removed. Removable flasks are called snap flask or "pop off" flasks. Popoff is a registered trademark. Some people make no distinction between the brand name and the type of flask.

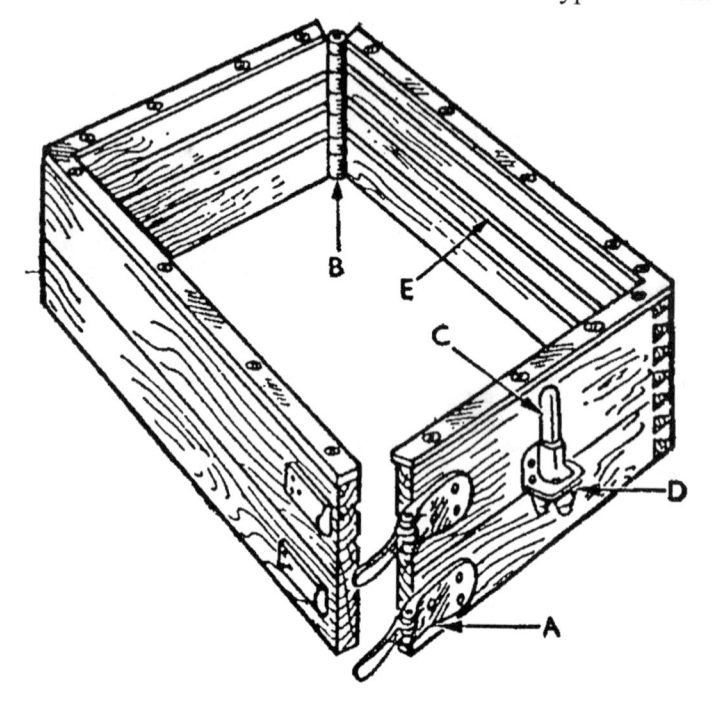

Parts of a Snap Flask: A. Lock, B. Hinge, C. Alignment Pin D. Pin Guide, E. Sand Retaining Grooves

Pins on the sides of the flask align the cope and drag with each other. Flasks may have single pins per side, double pins per side, angle irons or wooden wedges depending on the type of flask. If a flask is not tall enough for a particular casting, strips of wood or metal may be added to increase the height of the cope or drag. These strips are called "upsets" and are seen in the photo and drawing below.

This pop-off flask has an upset bolted to the top to provide additional cope height. It also uses double guide pins per side. The double guide pins help prevent the cope from rocking when lifting.

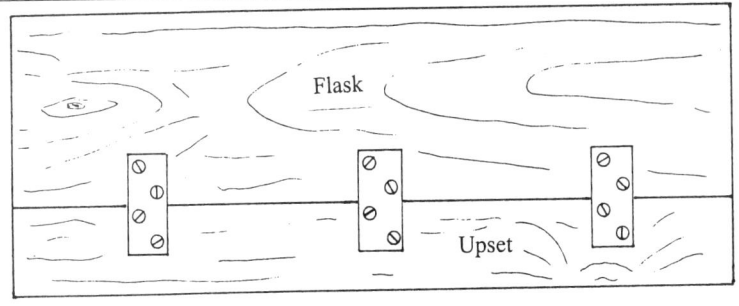

Wood Strips may be attached to flask with metal strips. They may or may not be screwed into the inside surface of the flask for a temporary situation.

11

Flasks do not have to be square, they may be round, hexagonal, or have an unusual parting line between the cope and drag as seen in the following photos.

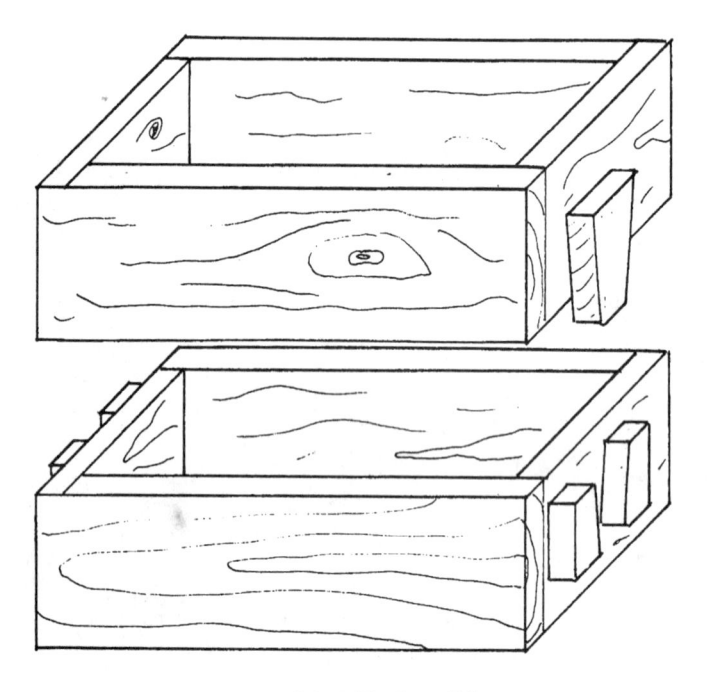

Wood Flask with Wedge Pins

Removable or snap flasks save a lot of space in your foundry because one flask can produce an infinite number of molds. A few sizes of removable flasks are the best way to go if you are planning to pour more than a few parts.

Commercial flasks are very expensive. You may be able to purchase a few discarded flasks that only need cleaning and aligning. Check around at foundries in your area, especially if they have changed from green to air-set sand. They may have several discarded flasks available.

You can make good quality flasks in your home shop. For simple molds, wedge type pins will work. When your projects require straight draws of plate mounted patterns

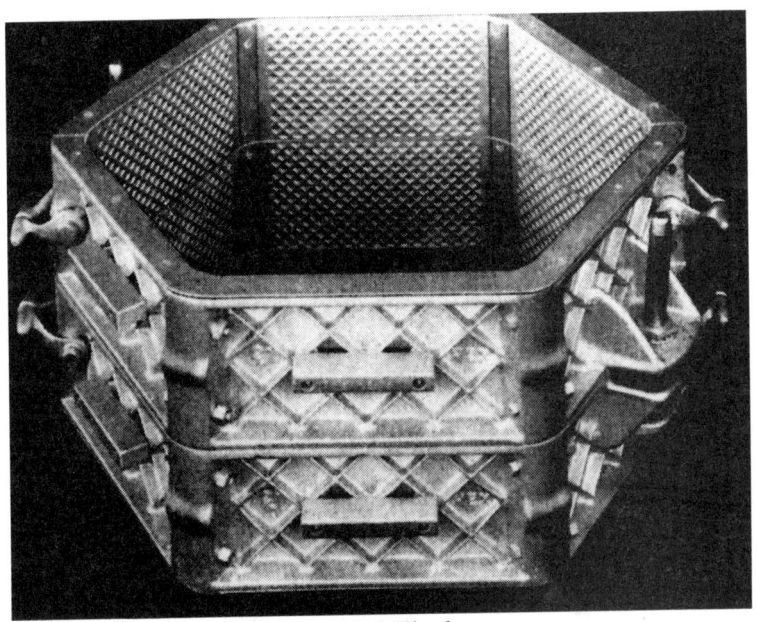

Hexagonal Flask

Round Flasks with Cheeks

Notice the holes cut into the sides of the flasks to vent steam and gas, also notice the pouring basins on top. The basins help prevent the aspiration of air.

(match plates) then the pins become a critical part of the flask. The pins are accurately aligned and mounted on the drag of match plate flasks.

This 18 x 20-inch flask is made of wood. It has eight 3/8-inch threaded rods reinforcing it so that it does not come apart from ramming. Notice the sand retaining grooves cut into the inside surfaces.

I built my first several flasks from wood. The butt-jointed flasks have served well but loosened up from ramming or burned up from spilled metal. Finger joints are best for making flasks. If you must use butt joints, they must be glued and screwed together. Threaded rod is also good reinforcing material for larger flasks. Casting several flasks and machining the top and bottom surfaces flat is also a good project for a small foundry. If you are going to build a flask, make it very sturdy and it will produce good molds for a long time.

Jacket: After taking off a removable or snap flask, a **jacket** slips over the sand mold to prevent runouts or molten metal leaks. They are usually made of metal but some are made of wood. Jackets are moved from mold to mold after each one is poured so a few jackets will cover many molds. Some foundries don't use jackets and let the bad molds leak. You may not need jackets for aluminum work, but iron and bronze molds will often leak from the pressure of the flowing metal if they are not jacketed.

Sand mold from a snap flask with an aluminum jacket. The mold has an iron weight on top to prevent the cope from floating and it has a shop made aluminum bottom board.

The **Bottom board** goes on top of the rammed drag before it is turned over. The bottom board remains with the sand mold until the casting is removed. The bottom board is also made of heavy plywood and is reinforced with wooden 2 x 4's or 4 x 4's. The 2 x 4's also raise the mold up off the ground so that you may get your fingers under it to lift the mold.

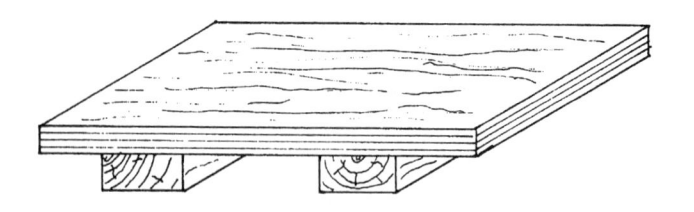

The **Molding board** or **Follow board** is a board or plate that holds the pattern while the drag is being rammed. For small molds you may use a section of ¾-inch plywood, however for larger or critical molds where absolute rigidity is needed, two sections of plywood may be fastened together to make a 1 ½-inch thick board. The molding board looks like the bottom board in most situations. When a flask is going to be used with a match plate, the pins are on the drag so that the cope and the match plate may be lifted straight up. In such cases the molding board must have several layers of wooden strips to provide clearance for the pins.

This molding board has additional height for pin clearance

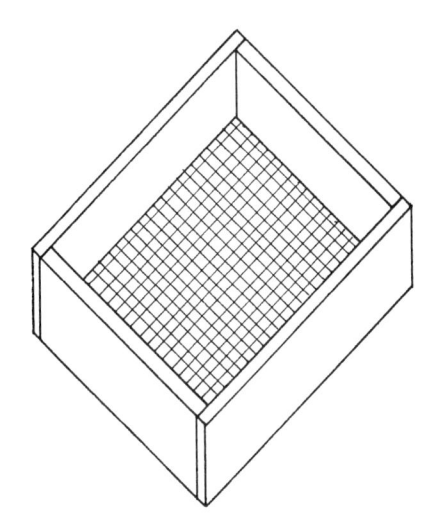

Riddle: Sand is screened through a **riddle** to remove trash and lumps before it is poured into the flask. Stretching hardware cloth with ¼-inch holes over a wood frame makes a small hand riddle.

Molding Tools: Commercial molding tools are available, but are not necessary to make sand molds. You may use small trowels and spoons. The Y shaped rapping bar in the lower photo (next page) is cast in brass but may also be made from a U shaped section of rebar with a handle welded on. The rapping spike is a section of 3/8-inch diameter rod with a point ground on one end. Dry wall screws and screw eyes may be used to grip and lift a pattern from the sand. Commercial rammers are available, but a beginner may make one from a section of yellow pine. The rammer in the lower photo (next page) is cast from scrap aluminum.

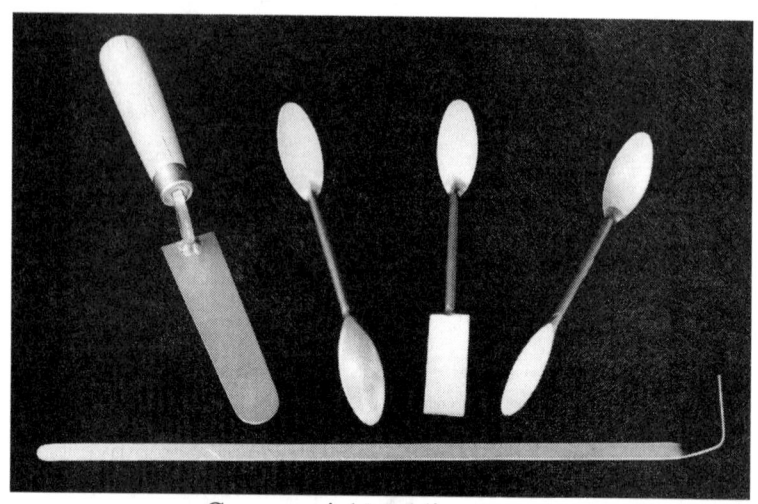

Commercial Molding Tools

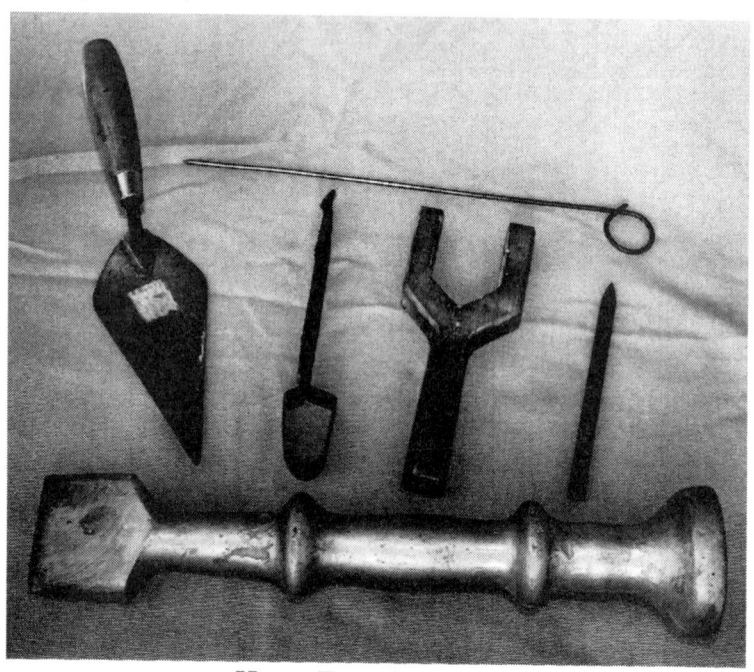

Home Foundry Tools

For small molds you can use a hammer handle as a rammer. A 10 to 12-inch section of ½-inch diameter steel tubing makes a good sprue cutter. Riser cutters are made from sections of steel tubing of various diameters that may be picked up as scrap from muffler shops.

Parting dust prevents the molding sand from sticking to the pattern. It forms a dry barrier between the cope and the drag so that they are easily separated. Parting dust must be a non-absorbent dry powder. Commercial parting dust is a plastic powder ground to a flour like consistency. You may be able to purchase a few pounds from a local foundry. You may also find it in small quantities from Budget Casting Supply among others in the suppliers list located in the appendix. Pumice and graphite may be used when proper parting dust is not available. Do not use silica flour for parting because it is known to cause silicosis. A sock makes a good dusting bag for parting compound.

Bending a loop in a section of stiff wire salvaged from a paper local election sign forms the vent rod seen in the lower photo. The wire is about 1/8-inch diameter with a point ground on one end.

Compressed air is used for blowing off the mold and removing loose sand. Some prefer to use a fireplace bellows or molder's bellows.

Metal Handling Tools:

Welding a short section of angle iron to a 3-foot long section of 3/8-inch rod or small rebar makes a skimmer. The angle iron is ground to match the inside surface of the crucible. Drill a few holes in the angle to allow metal to flow through the skimmer back into the crucible. The skimmer should be coated with a refractory wash and dried if you are working in aluminum and concerned about iron contamination causing brittle castings. Any tools that contact molten metal must be preheated before use.

Crucible tongs and pouring shanks can be easily forged from strap iron or purchased from various suppliers listed in the appendix.

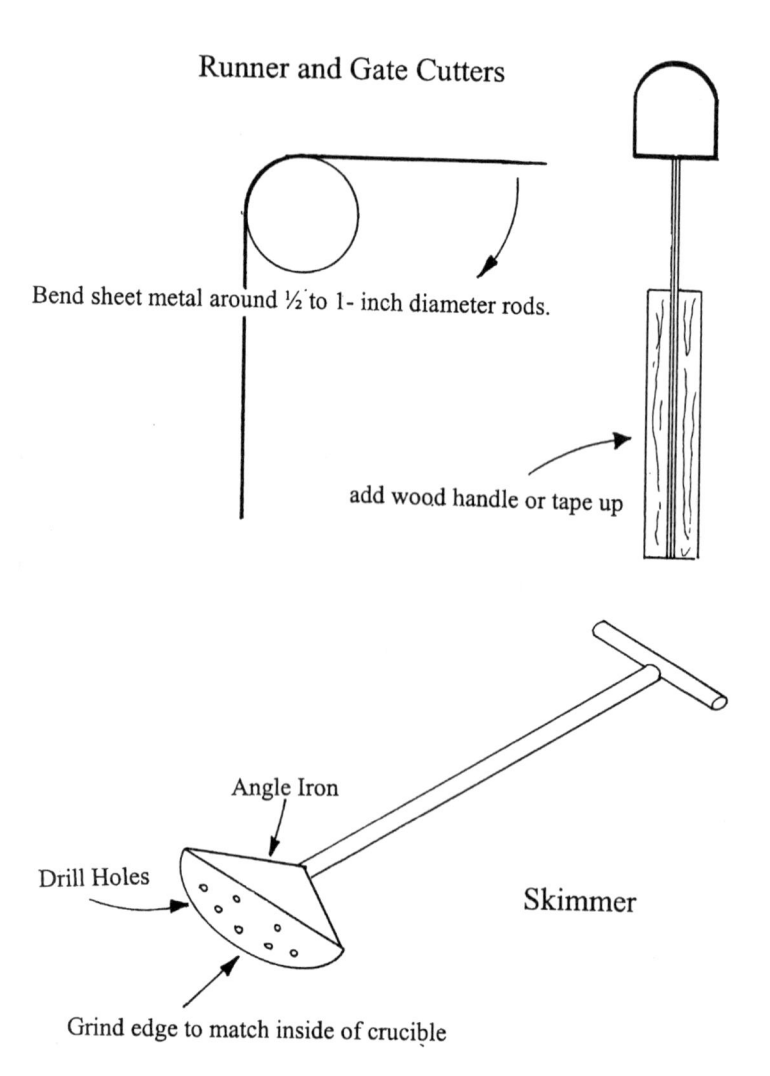

Runner and Gate Cutters

Bend sheet metal around ½ to 1-inch diameter rods.

add wood handle or tape up

Angle Iron

Drill Holes

Skimmer

Grind edge to match inside of crucible

II. FOUNDRY PROJECTS:

Because the match plate vibrator requires construction of a core box, split patterns, cope casting, and a cast in metal insert, it is a good example casting project. You may select a more simple first casting project, however this project is not difficult to cast and is essential for the match plate flask project found later in the book.

Make a Match Plate Vibrator:

Board mounted patterns require a vibrator to release from the sand. There are many types of vibrators including electric, pneumatic piston and pneumatic ball types. While piston type vibrators are popular, they require springs, close tolerances and must be oiled frequently. Ball type vibrators are very inexpensive and easy to make. They accept wide tolerances and may be finished on a disk sander making them an excellent small foundry project.

The ball type vibrator uses compressed air to spin a single ball bearing in a circular path. The spinning ball causes a rotating unbalance with a force of:

$F = W v^2 / g r$

$F =$ Force in pounds
$W =$ Weight of ball in pounds
$v =$ velocity - speed in feet per second if ball was traveling in straight line.
$g =$ force of gravity – 32 feet/second^2
$r =$ radius of circular path

Because the velocity term is squared, vibration increases rapidly with increasing speed of the ball.

The project is sized for a 5/8-inch ball bearing that should work for most smaller match plates. It is easily modified for both smaller and larger bearings.

The ball runs in a steel insert made from a section of 2-inch black iron pipe. A ball running on an aluminum surface runs the risk of rupturing a brittle casting causing a potentially dangerous or catastrophic failure. A steel insert is under compressive stress from the contraction of the aluminum casting around it, raising its strength similar to shrinking steel sleeves around gun barrels.

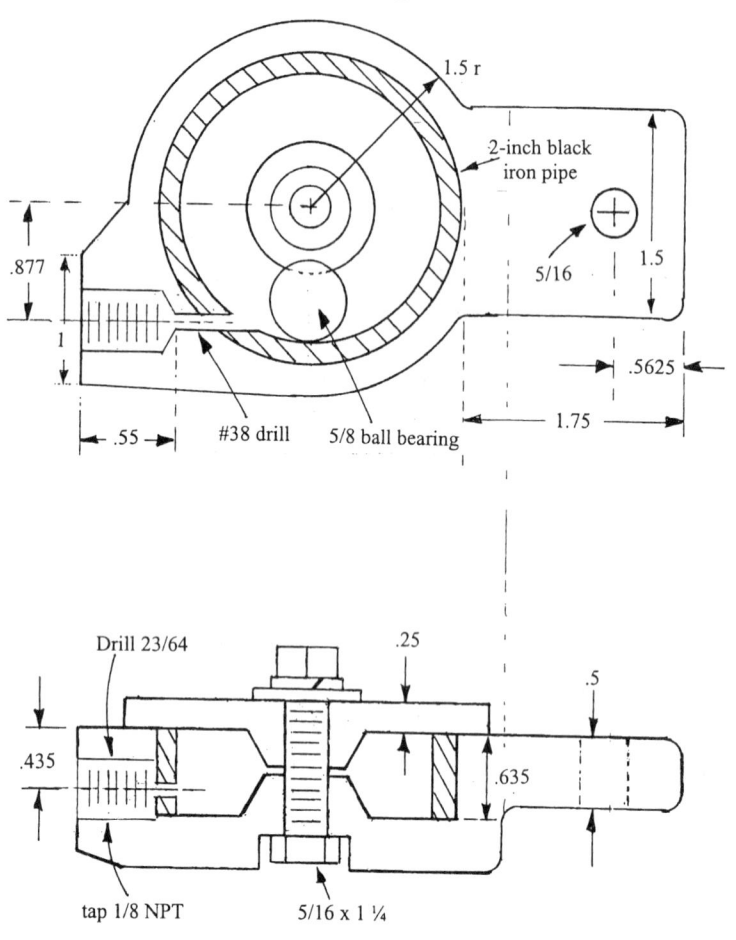

Match Plate Vibrator

Making the Patterns:

Draw the top view of the vibrator and make a photo copy in case you make a mistake and need another. Cut out the copied drawing and glue it to a 1-inch thick section of yellow pine. You may use other types of wood, but yellow pine is very inexpensive and a good hard wood that works well. I purchase 2-inch x 6-inch x 8-foot long yellow pine for much of my pattern work and saw or plane it to thickness.

When the glue on the drawing has dried, cut out the pattern on a band saw. Round the edges with a sander and fill all the rough spots with Bondo or similar auto body filler. It is easy to sand the draft or taper to the sides of the pattern by using a drum sander chucked in a small drill press, and tilting the table 2 degrees. Give the pattern 2 coats of shellac, sanding between the coats.

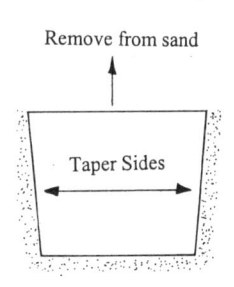

Center punch and using a small drill (1/8-inch) drill a hole through the pattern. Turn the pattern over and drill a .75 inch diameter hole 0.265-inches deep. Later, using Bondo, fill the bottom as needed to give a flat surface. This hole accepts the bolt head. Sand or bore a 2-degree taper on the hole.

Turn the pattern over. Using a number D drill (.246-inch), drill a ½-inch deep hole at the center of the pattern to mount the alignment pin. Cut a 5/8-inch length of ¼-inch diameter brass rod (or dowel) and round the edges.

Put glue into the hole and drive the dowel down until approximately 0.175-inch protrudes from the surface. Set the pattern aside to dry.

The Core Box: The core box is made on a metal lathe, however a wood lathe will also work.

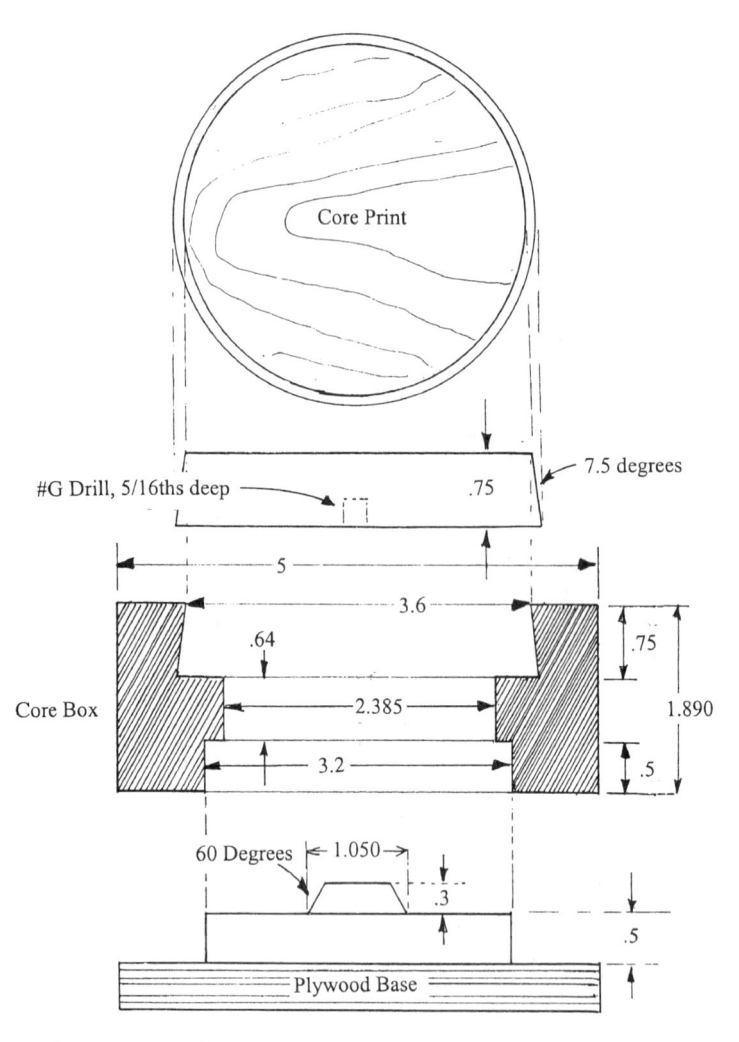

Cut two 12-inch long strips 2 ½ -inch wide from the 1-inch thick yellow pine stock. Glue and clamp them together making a 2-inch thick strip. When the glue has dried, plane the sides flat and saw the board in half (two 6-inch lengths). Clamp the halves together and make another cut so that the blocks are exactly the same length and square. Select two close fitting edges for the parting line. Drive two brads or small nails into one of the parting surfaces as seen in the photo. Be sure that the brads are perpendicular to the surface and not bent at an angle. Using a hand grinder or

Completed Core Box (open) with Steel Insert

sturdy snips, cut the brads off and grind or file them until they protrude 0.1-inch from the surface. Round the corners of the nails. Carefully set the mating surface against the nails, being careful to keep things square. Rap the back of the wood block with a hammer to mark the location of the holes on the mating piece of wood. Measure the diameter of the brads and drill slightly larger diameter (a few thousandths of an inch) sized holes at the locations marked. Transfer the blocks to the lathe.

Face one side square, drill and bore a 2.385-inch diameter hole through the center point of the parting line. Using a shop-vac to catch the wood chips at the tool bit makes this job much more pleasant. This hole centers the steel pipe insert to the core.

The remaining holes are cut to a convenient diameter and may be altered to suit materials on hand. Mark the

Boring Core Box while a Shop-Vac Collects Chips

depth of the next hole by measuring in 0.5-inches (deep) and placing the point of the boring bar at this location. Turn the lathe on and make a small cut. Bore a 3.2-inch diameter hole to this mark to accept the base of the core mold. It is important to cut this hole before removing the core box from the lathe because the base and the steel ring must be on the same axis (must line up). The base must be centered with the steel ring.

Remove the core box from the lathe, turn the box over, square and center it in the chuck. Face the box to a total length of 1.89-inches. Bore a 3.6-inch diameter hole 0.75 inches deep. Taper it 7.5 degrees as seen in the drawing. It is less critical if this hole is *slightly* off center because there is a little clearance in the wall thickness of the casting.

The center hole of the core box should now be 0.64 inches deep. Check the dimensions of the remaining holes against the drawing.

Sand the surface of the core box and give it two coats of shellac. Sand the inside smooth.

While the lathe is still set for the 7 ½ degree taper, mount and turn the core print from a section of 0.75 thick yellow pine. Drill the center hole using a #G drill.

Mount and turn two disks (left). One for the top and one for the bottom of the vibrator. One disk should be 3.2-inches in diameter and 0.8-inches thick. Face off material until you turn a cylinder 1.05 inches in diameter and 0.3 inches long in the center of the disk. Set the tool post for the 60-degree angle and cut the angle on the center post. Sand the disk and test fit the core box to it. Sand the disk until the core box easily fits over it but has little excess clearance. Remove the disk and give it two coats of shellac. Sand until smooth and check the fit again. Mount the disk to a square section of plywood to form a removable cover for the core box.

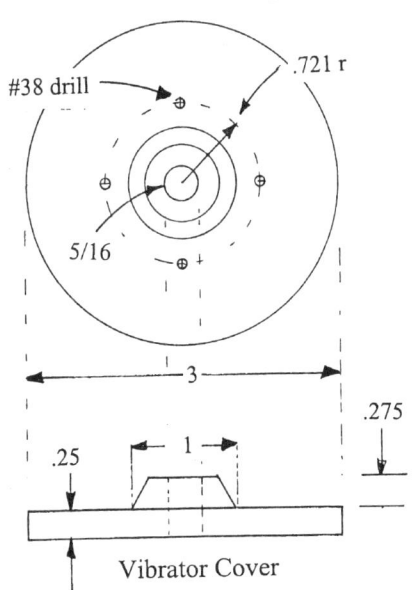

The vibrator cover is similar to the core box disc with the modified dimensions given in the drawing. The center post is made a little smaller. When you rap the pattern to remove it from the sand, the center post becomes somewhat enlarged. Extra clearance is needed in the pattern to accommodate this expansion.

The Steel Insert: The insert is made from a 0.635 inch long section of 2-inch diameter black pipe. The outside surface should be roughened to provide a better surface for the aluminum to grip. A few light grooves cut in the lathe or scratches made with a file or hacksaw are fine. The inside of the pipe should be smooth. You may smooth the weld by filing and sanding, however I run a boring bar through it taking a light cut. I make several rings at a time so boring is fast and easy. Part off the rings as needed or saw, grind and file them to length if you do not have a metal lathe. Remove the internal burrs so that you do not cut your self when handling the rings.

Making the Core: The core is made from a simple sand, flour and molasses mixture. Mix 1 part molasses with 10 parts hot water in a clean plastic gallon jug. You can store this in the refrigerator almost indefinitely so I usually mix at least a half a gallon at a time.

Using a section of rebar (one of my favorite foundry tools) or a strong expendable spoon, dry mix 2 pounds of clear fine sand with ¼ pound of wheat flour (the kind you buy at the grocery store). Be sure that all of the flour balls are broken up. Slowly add a little molasses water to the dry mix and stir it well with the rebar. Keep adding water in small amounts followed by vigorous stirring until a bond starts to develop. Don't get it too wet. It may take a few minutes for the flour to absorb the water and the bond to develop. The mass should hold its shape when squeezed but not be gooey or

Completed Core

watery. If little green strength develops, lightly dust flour over the sand and remix. Soon, you will develop a feel for the mixture and quickly get it right on subsequent batches.

On a flat surface, assemble the core box and clamp the sides together. Insert the steel ring and dust the interior of the box with parting dust. Partially fill the box with core mix and using a small rod or pencil, pack it into the corners around the ring. Fill the remainder of the box, packing the corners with the rod. Using a strip of sheet metal, smooth the mixture with a wiping rather than a cutting motion. The mixture tears and does not produce a smooth cut. When the box is full and the bottom is smooth, cover it with a section of thin sheet metal and invert it on the core plate (or baking sheet). Slide the sheet metal from under the core box and remove the clamp. Tap the box lightly and carefully open it.

Bake the core at 350 degrees for 35 to 40 minutes, periodically opening the oven to vent the steam. If you use a baking sheet, pay attention that the sheet does not warp when it gets hot because this will distort the core. When the core is finished, allow it to sit in a warm oven (200 degrees) until it is ready for use.

Making the Mold: You will need the basic home foundry tools and at least an 8-inch by 10-inch wedge pin flask. The rammer may be a hammer handle or can be cut from a 2 x 4. Later, you can make a proper pattern and pour an aluminum rammer. There are several good molding sand formulas in the sand chapter or you can use 100 pounds fine sand, approximately 120-150 mesh, and 7 pounds of bentonite and 3 pounds of water. The sand gets better after it has been in use for a while and the clay is evenly distributed over the sand grains.

Set the drag on a molding board or insert a molding board between the cope and drag if you are using a flask as seen in the photo.

Place the vibrator cover and the core print on the molding board and dust them with parting dust.

Screen a shovel of molding sand through the riddle.

Press the sand firmly around the patterns to keep them from shifting while you ram the sand.

Ram the sand all around the edges of the flask to ensure a strong bond. This prevents the sand from falling out of the flask when you open it. Continue ramming around the pattern being sure that the sand around the edges of the pattern is well rammed. This gives hard and sharp clean edges when the pattern is withdrawn.

Continue filling the flask and begin ramming with a rubber mallet.

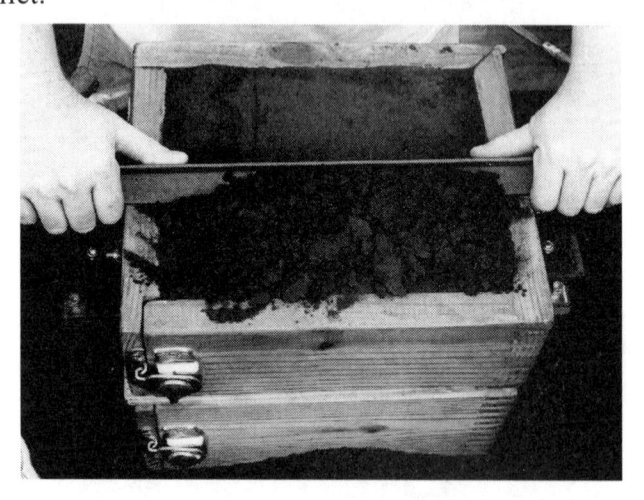

Using a section of angle iron, strike the surface flat. Fill in any low areas, ram and strike again.

Place a bottom board on the flask and roll it over. Remove the molding board. Place the vibrator pattern in place and dust with parting dust. Scribe a line with the vent rod to mark the location of the runner. This line will be transferred to the cope when it is filled. Replace the cope, ram until full and strike off as done with the drag.

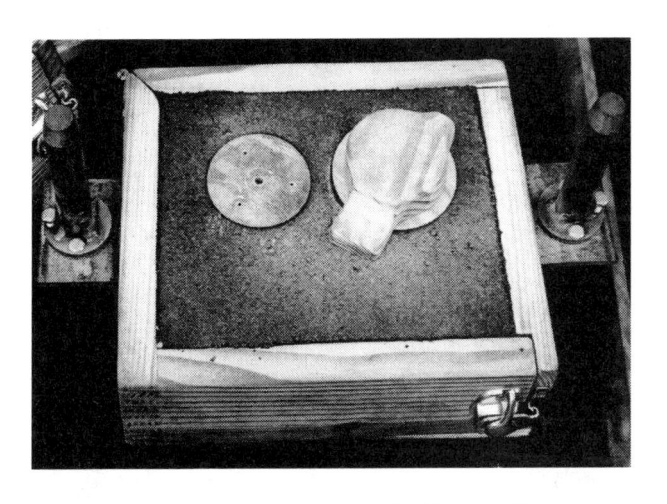

Drag Parting Surface with Vibrator Pattern

Open the mold and using a section of ½-inch tubing, cut a sprue.

Tap the spike into the pattern and rap in all directions to loosen the pattern from the sand. The motion is similar to ringing a small bell.

Using a screw-eye, or in this case a drywall screw, lift the pattern from the sand.

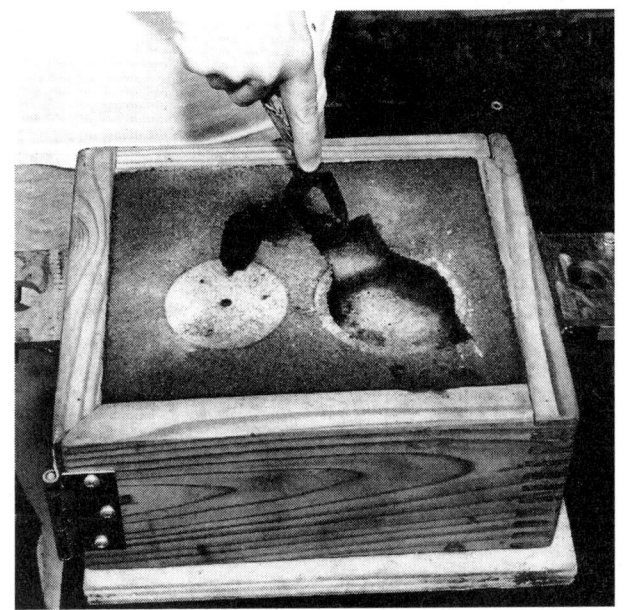

Cut the gates away from the pattern towards the sprue. Cut a small well in the drag under the sprue.

Vent the mold at the deepest locations and at the point farthest from the gate.

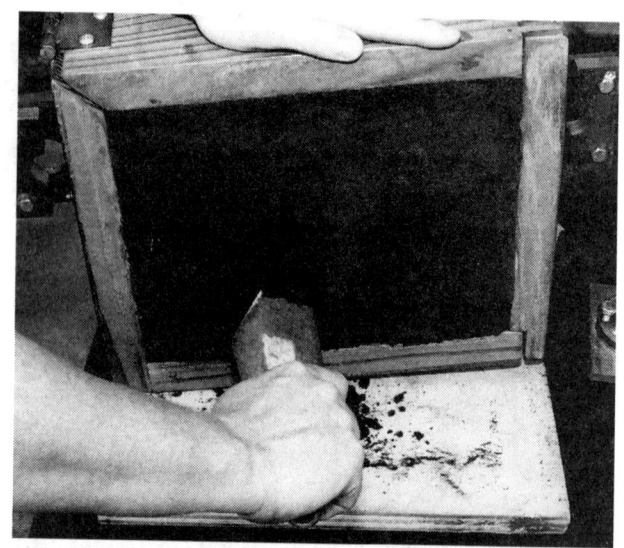

Using a small trowel cut a cone shaped entrance to the sprue. Smooth any loose sand with a spoon. Blow off the mold and blow out the sprue while running a finger through it to smooth any loose sand.

Insert the core. Turn the cope over BEFORE you get it over the mold so that any loose sand will not fall into the mold. Close the mold and if you are using a snap flask, remove the flask.

Completed Mold. Pouring the Casting.

Completed Casting

The mold is poured at 1325° F. Because the section thickness is the same throughout most of the casting and there is a metal chill at the thickest section, no risers are used.

Machining and Assembling the Vibrator: Saw off the gates and remove any fins on the casting with a drum sander. The top surface of the vibrator body casting is finished on a disk or belt sander. (Wear heavy leather gloves and be extremely careful. A sander will instantly remove the flesh right down to the bone if you contact the wheel with your hand.) You want to sand a flat surface around the insert. It does not have to be perfect, without pits. You only need to have a flat surface for the top to seat. Locate and drill the 5/16-inch center hole in the vibrator body.

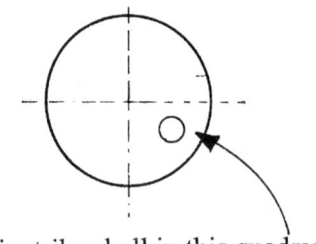

Air strikes ball in this quadrant

The only a critical machining operation is the location of the orifice. The air should strike the ball in the lower right hand side. The hole is located 0.877 from the center and 0.435 from the top edge. Drill the orifice by bolting the vibrator body to an angle plate. Center drill the location followed by a #18 drill 0.55 inches deep. Using very light pressure and a #38 drill, carefully drill through the steel sleeve. Use high speed and raise the drill frequently to clear chips. Wash the chips with a liberal application of WD-40 or equivalent light oil. Finish the hole by drilling with a 23/64-inch bit 0.55 inches deep. Tap 1/8 NPT. Install a 1/8 NPT to ¼-inch hose barb being careful not to over tighten the fitting and crack the case.

Center punch, drill the cover and the four vent holes using a #38 bit. Drill the center out to 5/16-inch. Attach the cover using a 5/16 x 1 ¼-inch bolt. Do not over tighten the cover so that it deflects and causes the ball to stick. If for some reason your cover is too low, you may shim it with a cardboard ring cut from the back of a pad of paper.

Ball Bearings may be found at hardware stores for approximately 50 cents each. They may be ordered from MSC Industrial Supply for approximately $13.50 per 100.

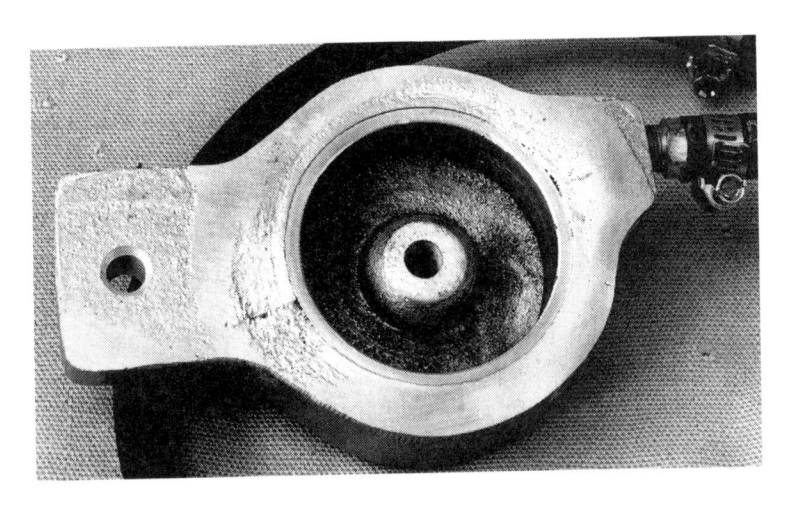

Completed Vibrator Body

Vibrator with Cover attached

Making Wooden Flasks: You can get by with a few butt jointed wedge pin flasks to start. An 8 x 10-inch and a 12 x 14-inch flask will cast many small projects and are quick to make. If you do much casting, you will soon find that the flask joints will loosen up or the flask will become burned up from spilled metal. As your projects become more exacting, you will want good dependable flasks. Finger jointed flasks do not loosen up and may be considered permanent. Good flasks are neither expensive nor difficult to build; however it does require much work. After you build a few good flasks, you will appreciate the several hundred-dollar price tag on a good commercially available flask.

Good flasks are made of hardwoods such as cherry. Yellow pine, being very inexpensive, makes good serviceable flasks. Because some new wood is not completely dried and may soon warp, it is good to purchase your wood and let it sit in your workshop for a few months. If you are just getting started, purchase some extra wood to dry and make a few butt jointed flasks to use now. By the time they loosen up, your wood supply will be ready to work.

Finger jointed flasks should be removable or snap-flasks so that you do not burn up all that hard work with one spill.

Flask hardware may be fabricated as seen in a few photos in this book or it may be cast. Door hinges can be used at the hinged corner, however they MUST NOT be attached with the counter-sunk screws that come with them. These screws will pull the flask out of alignment as they pull themselves into the hinges. Use pan head sheetmetal screws and assemble your flask on a flat surface such as the machined top of your table saw.

Good commercial case latches should be used if they are available. Strong cabinet latches will also be work, but do not hold up as well. Avoid the cheap pressed metal latches.

They will fail after two or three molds and are a complete waste of time.

The inside of the flasks should have two or three shallow grooves cut around the inside to hold the sand in place. Rounded grooves may be cut with a router or ball end mill before the flask is assembled. When setting up your jigs, it is good to cut several sized flasks at once to reduce the amount of labor. Machine the longest or largest sections first. If you make a mistake, you can cut it down and use it on a shorter flask.

It is not difficult to make finger joints on a table saw, however it does require making a wooden jig and a few test cuts. Be sure that the wood stock is squarely and securely clamped at all times. The setup for finger joints is seen below:

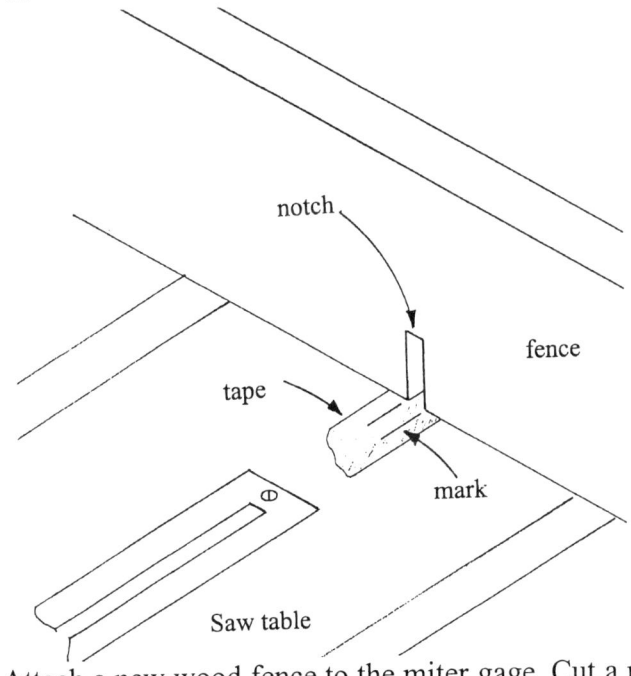

1. Attach a new wood fence to the miter gage. Cut a notch using the dado head set to ¼-inch. Mark the sides of the cut on a piece of masking tape attached to the saw table.

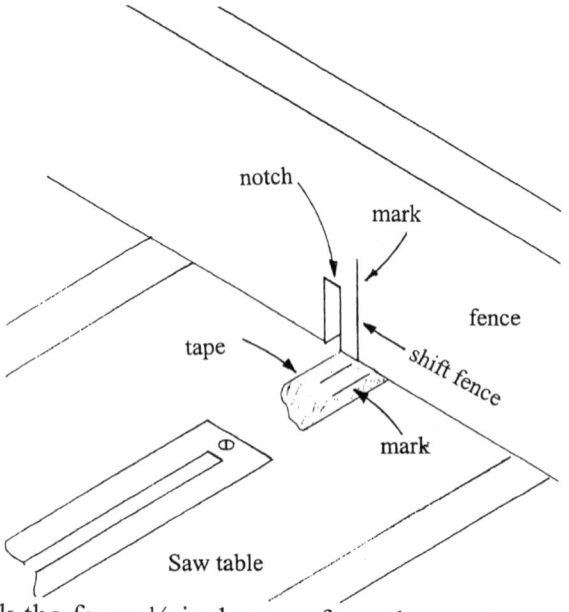

2. Mark the fence ¼-inch away from the notch. Using the mark, detach the fence and shift it ¼-inch to the right.

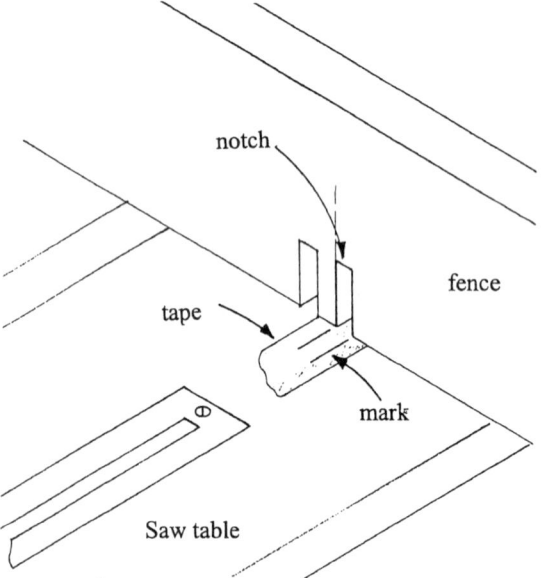

3. Cut a second notch in the fence.

Insert block and glue in place

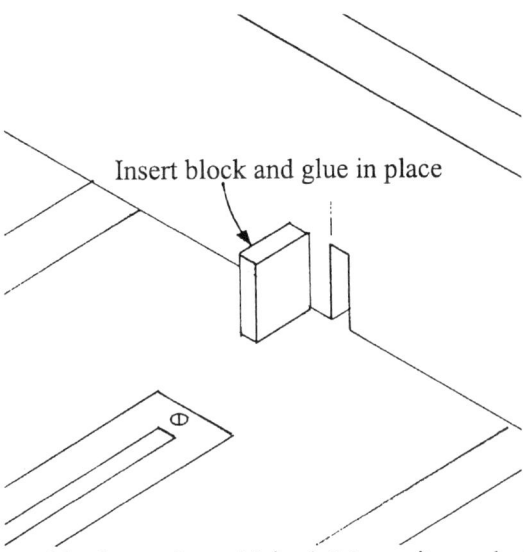

4. Insert a block cut from ¼-inch Masonite and glue it into the notch. The block must be the exact thickness of the notch.

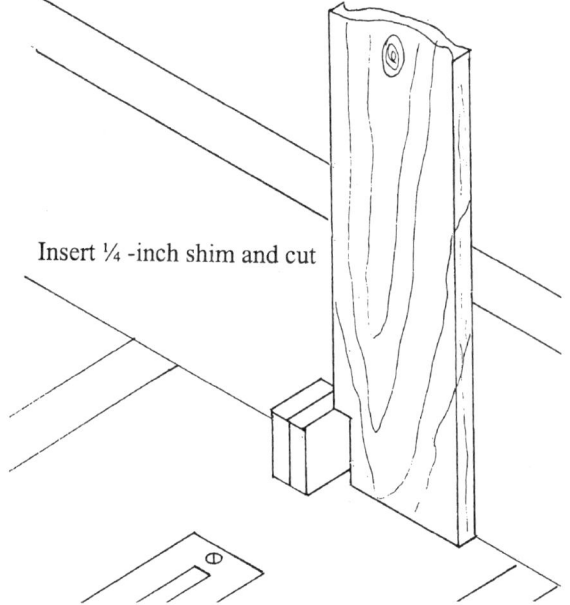

Insert ¼-inch shim and cut

5. Using a shim, notch the first board.

Remove shim, shift to pin and add second board

6. Add the second board after removing the shim and shifting the cut flush to the pin. Clamp the boards together and make a second cut.

7. Continue cutting and shifting the clamped boards.

8. Assemble the test sections to see that the pins fit and the tops are flush. If the pins do not fit, the width of the dado or the spacing of the pin in the fence is off.

9. Completed finger joint on yellow pine flask. Shop cast hardware is used on this flask.

Assemble the sides of the flask on a flat surface using a square. It is important to keep the sides square both vertically and horizontally so that the pins will fit well. If you can not get them to square vertically, the finger joints may have been cut at an angle. <u>It is best to cut two sides from scrap and assemble them to be sure all of your setups are good before actually cutting your flask.</u> You will have to shim the guides if the flask is not square.

The wood stock used on the flask is planed to 1 1/8-inch thickness to make it sturdy but not so heavy as 1 ½-inch material. This flask will last for years. This particular flask is finished with shellac, which is quick and easy. I also finish some of my flask with linseed oil cut with a little turpentine or paint thinner. I add a little japan drier to speed the curing process.

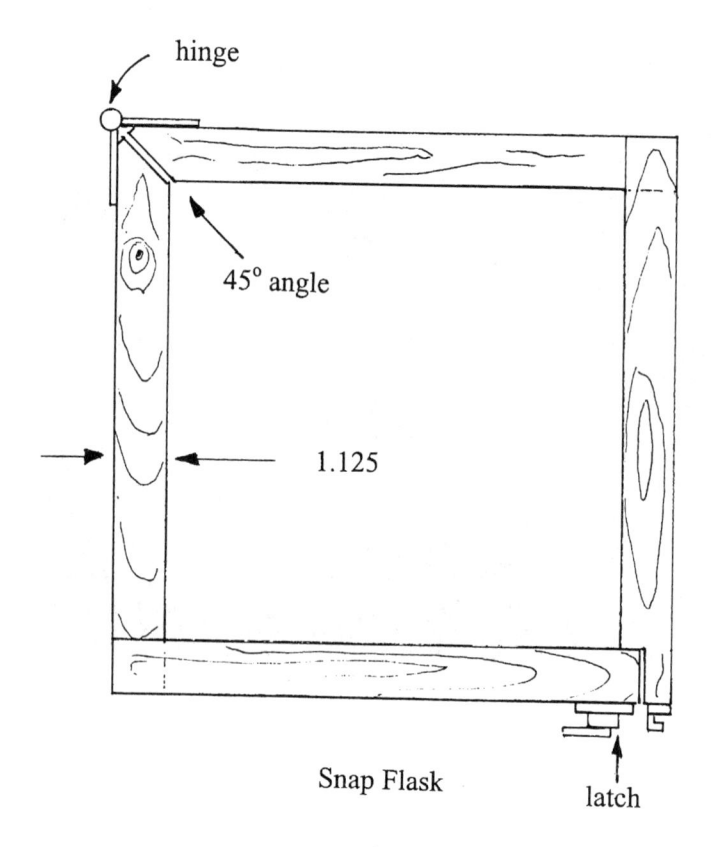

hinge

45° angle

1.125

Snap Flask   latch

Making Flask Hardware: Wooden pins are satisfactory for loose patterns because they are rapped to loosen them from the sand. If the pattern will not release, you can rap it until the clearances are great enough to remove it. Plate mounted patterns are not rapped to loosen them from the sand, they are vibrated and drawn straight up along the pins. The best match plate patterns have pin bushings similar to the cope that guide the pattern straight up as it is withdrawn from the mold. Deep patterns require a straight draw to prevent them from breaking the mold.

Cast flask hardware with vibrator installed on a match plate. Note the proper way to lift the cope using the pin.

The fabricated hardware seen on some of the flasks in this book is acceptable, however the cast aluminum hardware is superior and makes a good foundry project.

The hardware consists of 3 cast parts and one 1-inch diameter steel pin which make one pin set. Each flask requires 2 pin sets, one on each side. The patterns can be mounted to a match plate after the first flask has been made. You may pour two sets at one time using the match plate method.

The hardware is cast and little machining is required to finish it. The pin guides are cast against chills that form precision surfaces. The larger patterns are cut from yellow pine that has been planed to ½ inch thickness. In order to reduce the machining required, little or no draft is used on

the outside surface of the cope guide as noted in the drawing. Because you are rapping loose patterns, you can (usually)get them out of the sand.

The pin guides have precision surfaces to hold the metal chill. Make these by cutting a sufficient notch, approximately 1 1/8-inch wide and 1/2-inch deep. Fill the notch with Bondo auto body putty and square the pin guide to the miter slot in the saw table. Lightly coat a section of 1-inch diameter rod with petroleum jelly and carefully work the rod down through the Bondo to rest on a set of parallels as seen in the photo. In a few minutes, the rod may be removed and the excess trimmed with a wood chisel or single edge razorblade. This forms a precision surface to hold the metal chill. The pin guides will cast with a hole that is accurately located and has a smooth surface. No machining is required. The upper pin bushings are formed in a similar fashion.

Cut a 1-inch diameter rod to the exact length of the pattern to use as a chill. Heat the chill to approximately 150

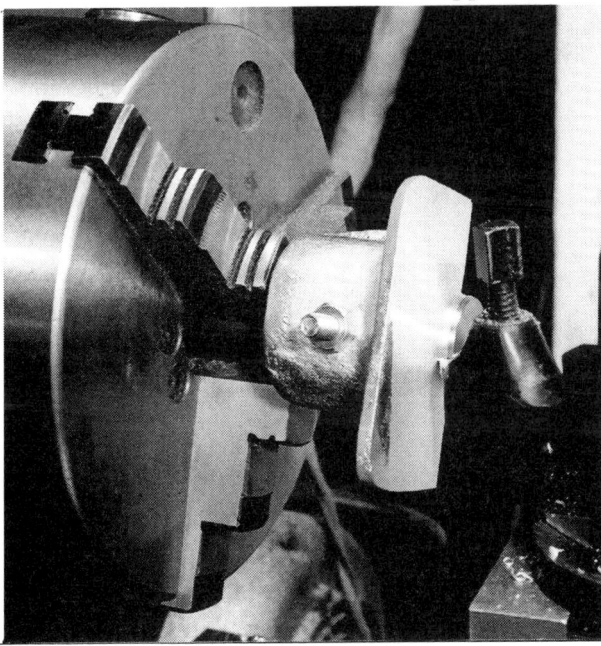

Finishing the cope pin bushing in a lathe. The pin bushing may also be finished on a disk sander.

49

degrees to prevent water from condensing on it while in the mold. Set the chill into the precision channel formed in the pin guide and ram up the mold. When the pattern is withdrawn, the chill remains in the sand. Close the mold and pour. Knock the chill out using a hammer when the casting is cool.

Make the cope pin bushings from the same material as the drag pin guides. Use ¼-inch Masonite for the base of the bushing. Make a generous fillet between the base and the body of the bushing. Ram two of them in the drag, turn the mold over and cut a relief in the sand between them to accept the chill. Gate into the heavy section with the thin sections on the outside (see drawing). The bushings may be finished on a sander or on a lathe. Make a small arbor to

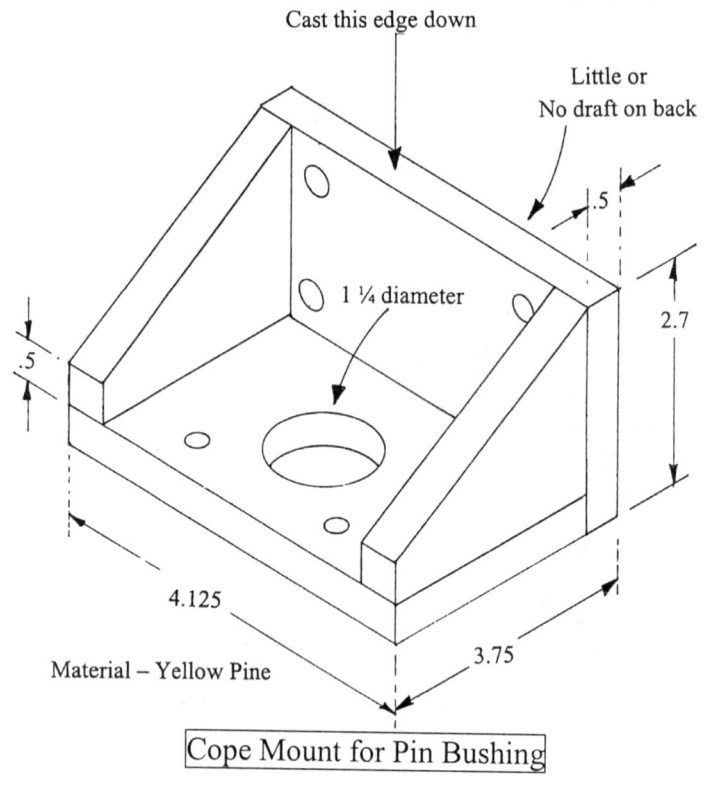

Cast this edge down

Little or No draft on back

1 ¼ diameter

.5

2.7

.5

4.125

3.75

Material – Yellow Pine

Cope Mount for Pin Bushing

hold the pin guide from a 4 to 6 inch length of 1-inch diameter steel rod. Drill a hole in the rod 9/32-inch diameter about ¾ to 7/8 of an inch back from one end.

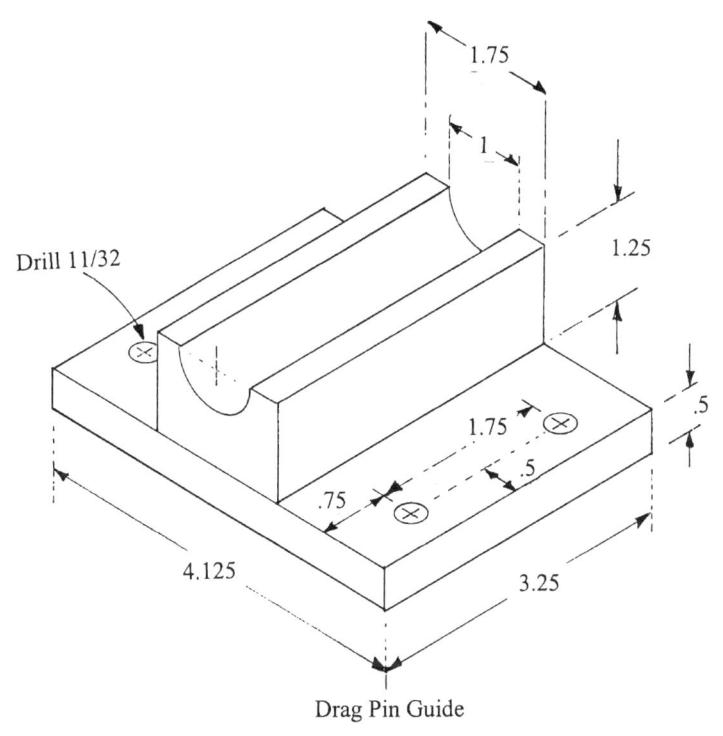

Drag Pin Guide

Drill the cope pin guide 9/32 and bolt it to the rod using a 2 ½-inch long ¼-inch bolt. The pin guide base may be cut square in a lathe or it may be sanded square by mounting the drag pin guide on the sander table to guide the pin squarely into the sander. I use 60 grit PSA disks and a crepe rubber block to clean the sanding

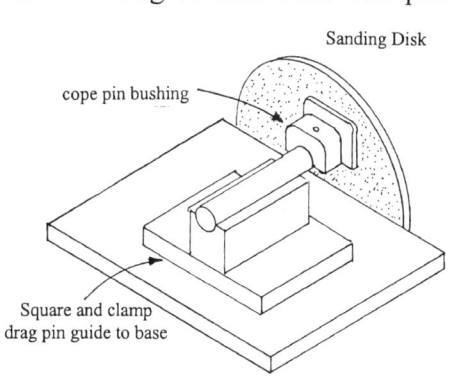

Sanding jig to square bottom of pin bushing

51

Pin Bushing -Cope

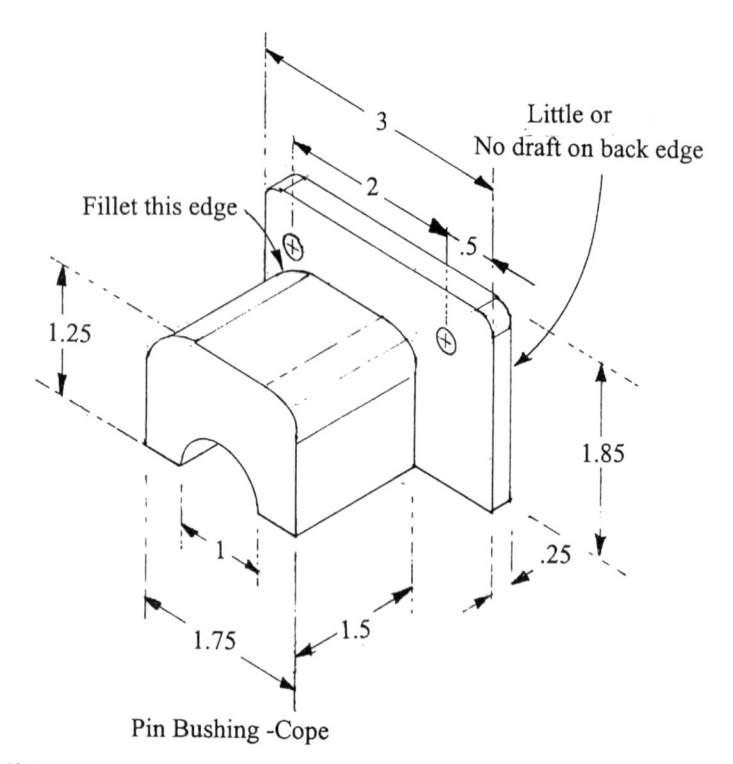

disk as necessary. Sanding is quick, but be careful because the parts become very hot and will probably require quenching before handling.

Round the front corners of the pin bushing bases so that they match those rounded corners on the rear. Slide the pin bushings on the cope mounts to check for high spots and clean up the cope mounts as necessary. Drill the pin bushing 9/32 to accept ¼-inch bolts. Attach the pin bushings to the cope mount with ¼-inch by 1 ¼-inch bolts. The holes are marked on the cope mount after assembly and drilled 5/16 to allow for adjustment.

Cut the flask pins so that they are ¼-inch down from the top edge of the assembled flask cope when they are located in the drag pin guide. Taper the end 7.5°. Clamp the pin in the guide and drill both 11/32 at the same time. You may file or grind a flat on the pin to make a better seat for the nut and washer. Bolt the assembly together using a 5/16 bolt.

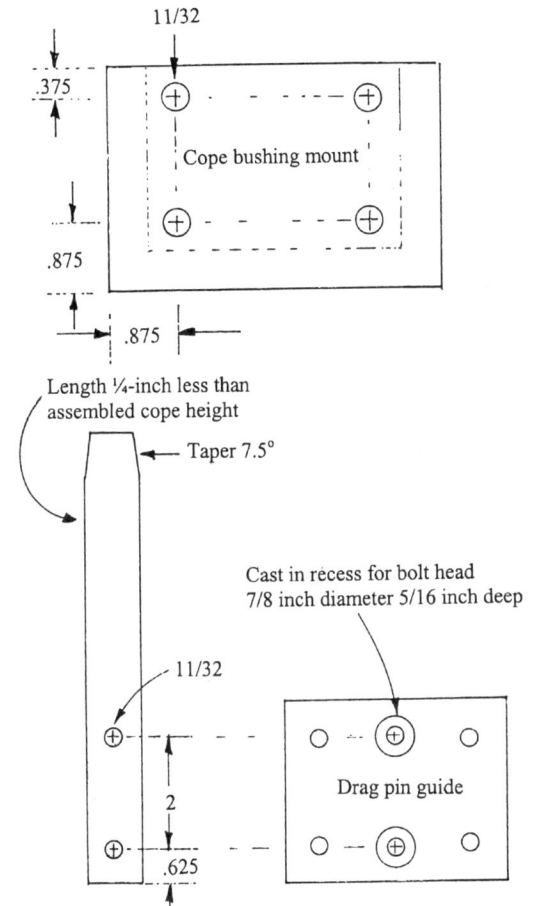

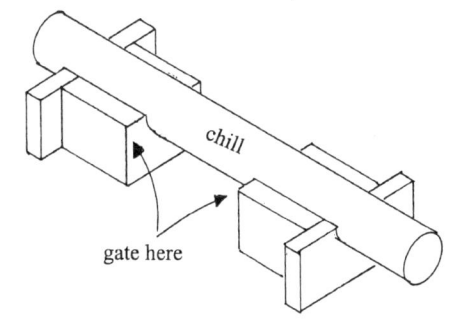

Cope bushing mount

Length ¼-inch less than assembled cope height

Taper 7.5°

Cast in recess for bolt head 7/8 inch diameter 5/16 inch deep

Drag pin guide

chill

gate here

To accept the bolt head and provide a flat surface for mounting, the pin guide has two recesses cast in the back ¾ to 7/8- inch diameter and 5/16-

inch deep. When mounting the hardware on your flask, the sides of the flask must be square or you will have to shim the guides.

Making the Aluminum Flask: The aluminum flask hardware is cast using a chill and finished as described above. Numbers and letters are added by gluing plastic letters found at office supply stores to a strip of cardboard cut from the back of a legal pad. The 14 indicates that it is a 14-inch "inside length" panel. Spare cast panels may be cut up and sections added or removed to make patterns of various lengths. Different lengths may be combined to form a wide variety of flasks.

Pour the mold at 1350 °F. Because the section thickness is fairly uniform and the thick section is chilled, risers are not used. Use four gates into one side and use small shrink bobs at each gate to prevent shrinkage at the joint between the casting and the gate. Because each flask requires

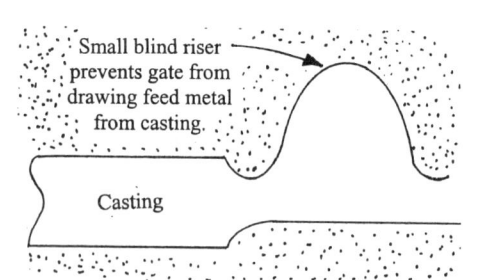

Small blind riser prevents gate from drawing feed metal from casting.

Casting

54

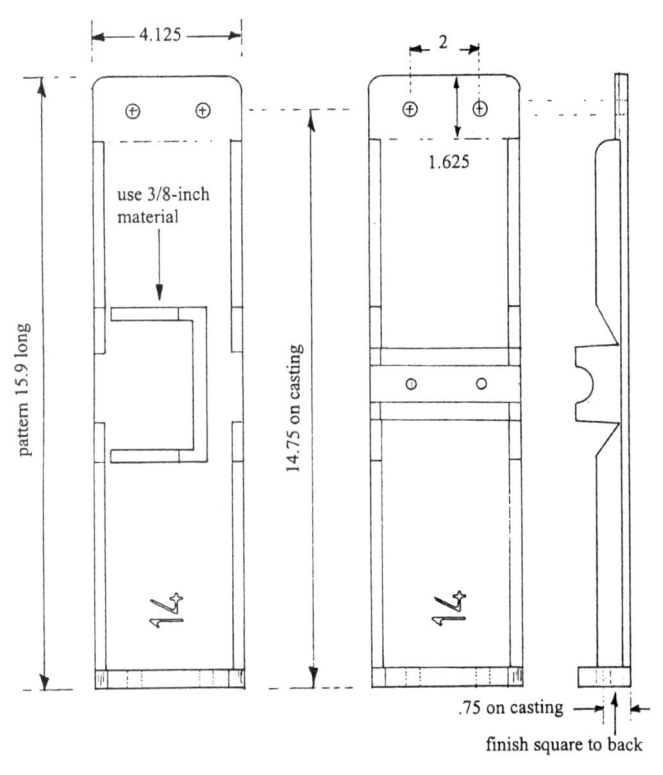

- 4.125
- 2
- 1.625
- pattern 15.9 long
- 14.75 on casting
- use 3/8-inch material
- .75 on casting
- finish square to back

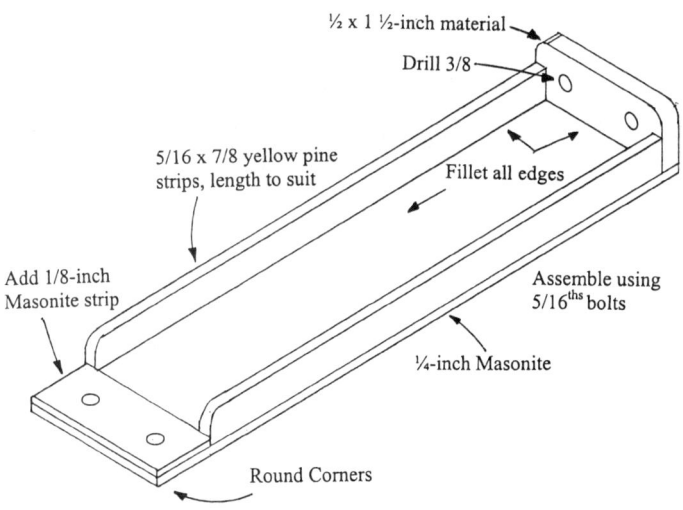

½ x 1 ½-inch material
Drill 3/8
5/16 x 7/8 yellow pine strips, length to suit
Fillet all edges
Add 1/8-inch Masonite strip
Assemble using 5/16ths bolts
¼-inch Masonite
Round Corners

eight panels, I mounted them to a match plate. Put one heavy section and one light section in each mold so that the metal requirements are similar. When panels are cast two at a time, the mold takes approximately 10 pounds of metal.

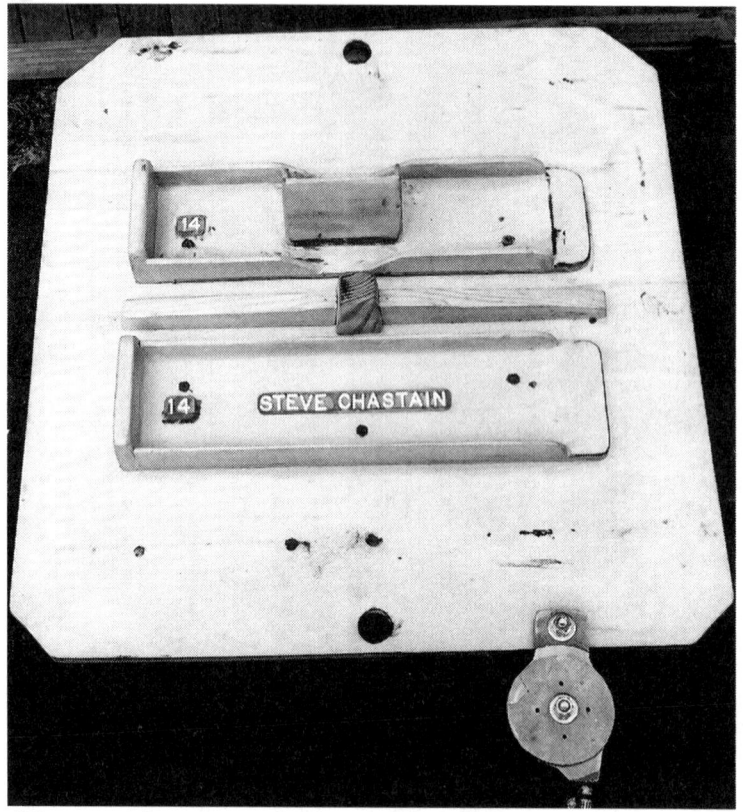

The gas furnace quickly melts this, however the 4-inch diameter pipe crucible must be at least 8 inches deep. The side arm must be reinforced to ½ inch diameter rod.

As seen in the drawing, make the panels from ¼-inch Masonite. A 1/8-inch strip is added to make a bolt boss at one end. Add draft to all edges. Square the mating surfaces after casting. The panels may be bolted to a mill table or squared using a fence on a sander. Assemble the flask on a flat surface such as the machined surface of a table saw. Shims may be necessary to square the flask.

Match Plate Casting with Chill in Lower Panel

Rammer:

Use a split pattern to cast a rammer. The pattern may be carved from two sections of wood, however my rammer is built up similar to a cannon pattern. Several scrap strips of yellow pine are assembled as seen in the drawing at the bottom of the next page. This pattern is quickly rounded either by sanding or on a lathe. The center grip is sized to be comfortable for my hand.

The finished rammer weighs approximately 3 pounds. With risers, the casting requires a full pot of metal. Finish the rammer on the sander.

Commercial Latches:

Patent drawings of snap flask hardware are seen at the end of the chapter. These are best cast in bronze and a project for the most ambitious new metal casters.

Next page: Upper Left: approximate dimensions for hand rammer. Upper right: Split pattern. Lower: Built up rammer pattern before finishing.

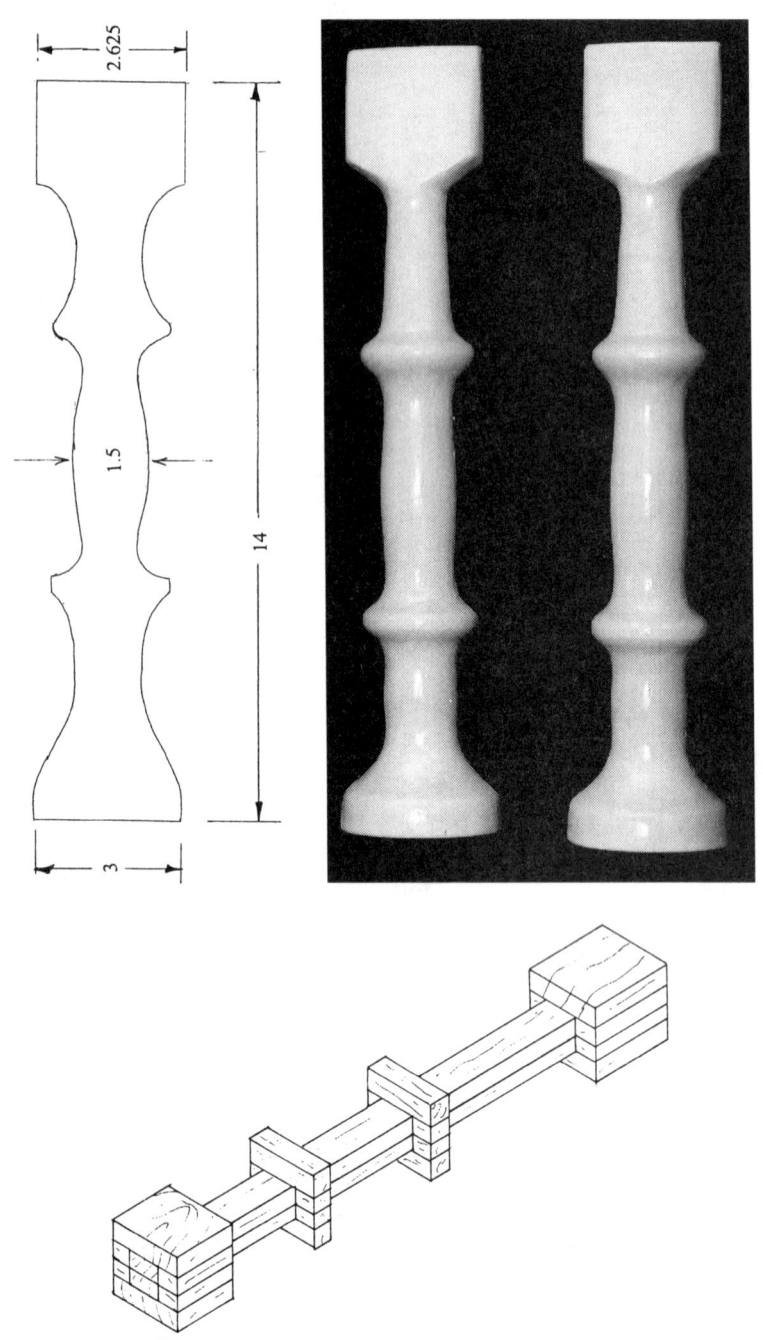

Rammer pattern made from built up sections.

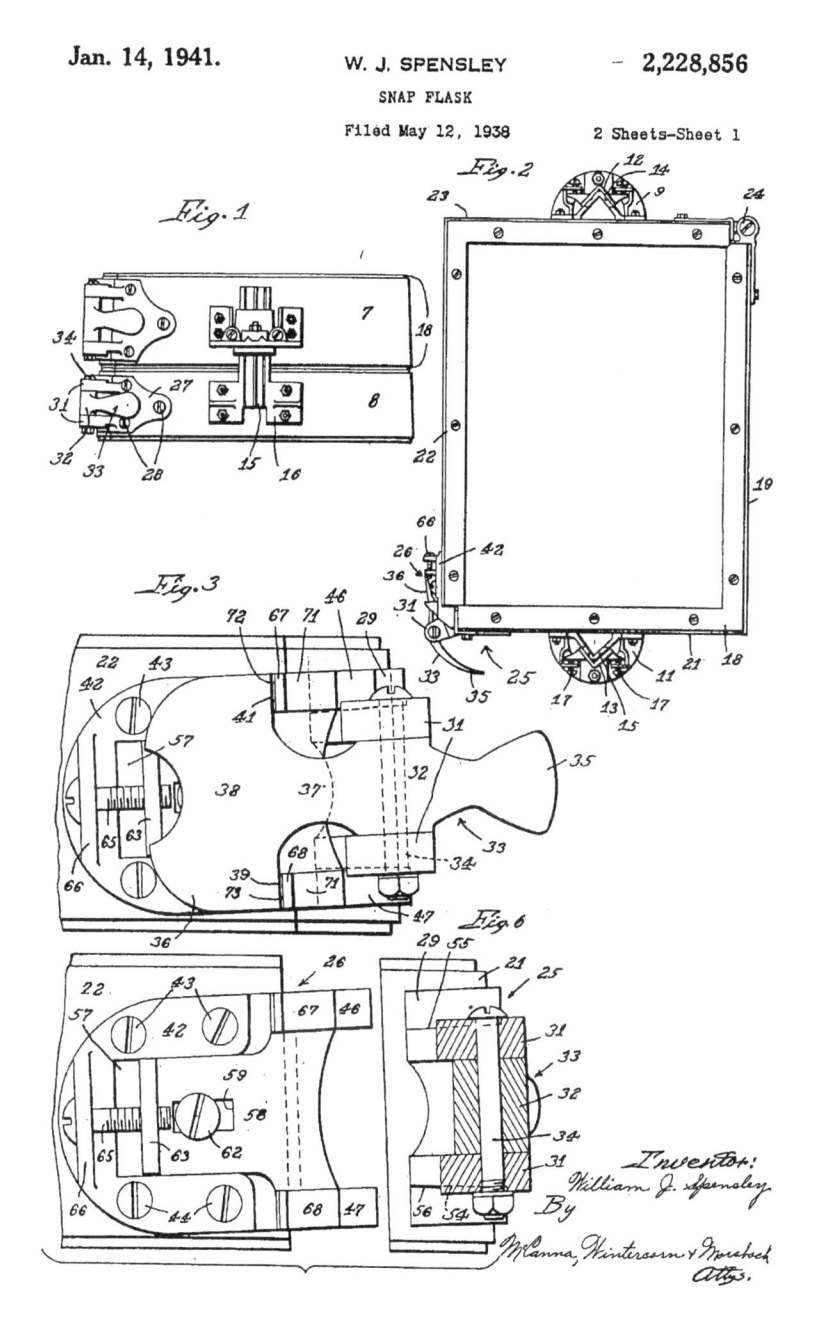

Typical Snap Flask Hardware

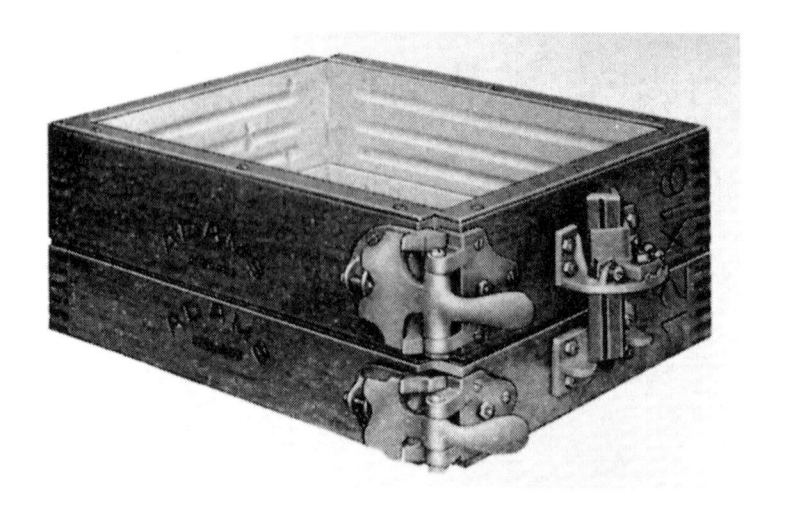

Jan. 14, 1941.　　　W. J. SPENSLEY　　　2,228,856
SNAP FLASK
Filed May 12, 1938　　　2 Sheets-Sheet 2

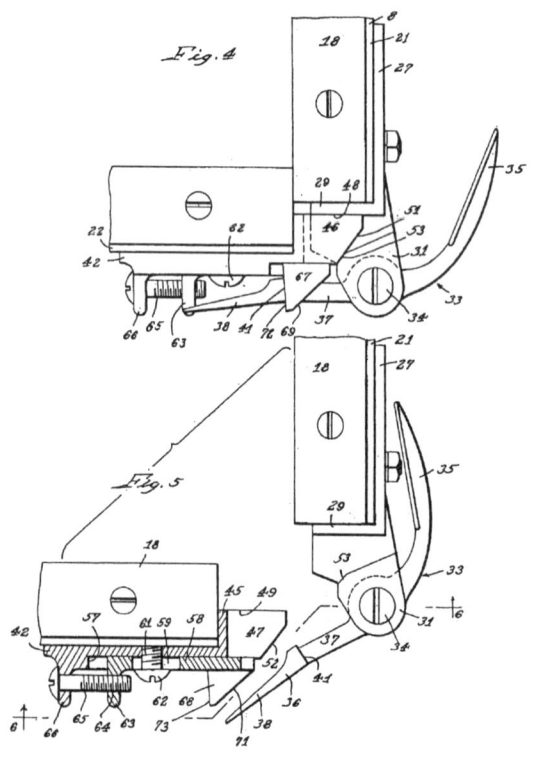

III. MELTING EQUIPMENT:

CRUCIBLES and LADLES:

Crucibles and ladles found in a small foundry are of the lip pouring type and the teapot or bottom pouring type. Tea pot type ladles are used to pour clean metal from underneath a layer of floating slag or flux.

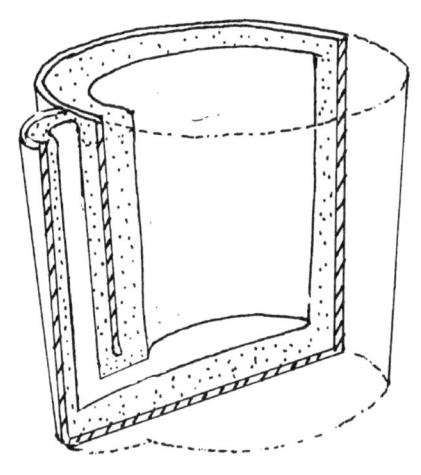

The most simple lip pouring crucible or ladle for a beginner's aluminum foundry is a 6-inch length of 4-inch diameter steel pipe with a disk cut from ¼-inch plate welded to the bottom (see the ladle in the beginner's small furnace section). A bare steel ladle will serve for many castings however steel dissolves in aluminum and the inside of the ladle should be coated to both protect the ladle and improve the quality of the aluminum.

Bottom Pouring Teapot Ladle

Commercially available crucibles may be made of a clay and graphite mixture called plumbago, or they may be made of silicon carbide. Plumbago crucibles are less expensive than silicon carbide, however silicon carbide crucibles are more durable.

Plumbago crucibles must be tempered before they are used for the first time. It must be slowly brought to a bright red heat and slowly cooled by leaving it in the furnace and letting the furnace cool back to room temperature. You may want to cover the exhaust port with a brick to slow the cooling cycle. Plumbago crucibles will absorb moisture from the air and must be tempered again if they are not stored in an airtight container. Because clay and graphite crucibles are delicate, you must use properly fitting tongs to avoid breaking them.

"Salamander" Clay-Graphite Crucible by Morganite

High temperature metals require a clay-graphite or silicon carbide crucible. The highest quality aluminum is melted in a crucible and not an iron pot. This prevents iron from dissolving into the aluminum and lowering its strength or electrical properties

American Standard Crucibles for Non-Tilting Furnaces

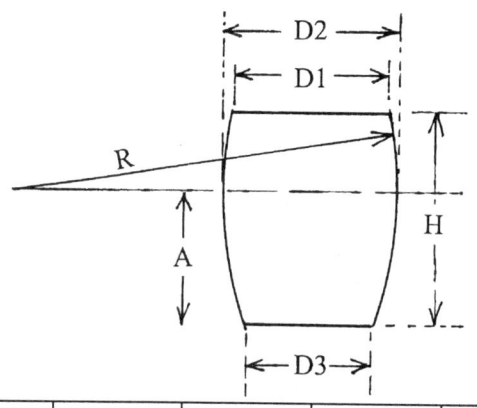

Size Number	H	A	D1	D2	D3
10	8 1/16	5 1/4	6 1/16	6 9/16	4 13/16
12	8 1/2	5 1/2	6 3/8	6 7/8	5 1/16
14	8 7/8	5 3/4	6 11/16	7 3/16	5 1/4
16	9 1/4	6	6 15/16	7 1/2	5 1/2
18	9 13/16	6 3/8	7 5/16	7 15/16	5 13/16
20	10 5/16	6 11/16	7 11/16	8 3/8	6 1/8
25	10 15/16	7 1/8	8 3/16	8 7/8	6 1/2
30	11 1/2	7 1/2	8 5/8	9 5/16	6 13/16
35	12	7 13/16	9	9 3/4	7 1/8
40	12 1/2	8 1/8	9 3/8	10 1/8	7 7/16
45	13 3/16	8 9/16	9 7/8	10 11/16	7 13/16
50	13 3/4	8 15/16	10 1/4	11 1/8	8 1/8
60	14 7/16	9 3/8	10 13/16	11 11/16	8 9/16
70	15 1/16	9 13/16	11 1/4	12 3/16	8 15/16
80	15 5/8	10 1/8	11 11/16	12 11/16	9 1/4
90	16 3/16	10 1/2	12 1/8	13 1/8	9 9/16
100	16 11/16	10 7/8	12 1/2	13 1/2	9 7/8
150	18 3/8	11 15/16	13 3/4	14 7/8	10 7/8
200	20	13	15	16 1/4	11 7/8

All dimensions in inches.

R is approximately .225 D2

A type Crucibles				
Size	Height	Top Diameter	Bottom Diameter	Capacity Brass
	inches	inches	inches	pounds
8	7 1/4	6 1/8	4 1/8	28
12	8 1/4	6 3/4	4 3/4	40
16	9 1/8	7 1/4	5 1/8	51
20	10 1/4	7 3/4	5 3/4	66
25	11	8 1/4	6 1/8	79
40	12 1/2	9 1/8	6 1/4	110
50	12 3/4	9 3/4	7 1/8	132
60	14 1/4	10 39/50	7 1/2	169
70	14 3/4	11 1/2	7 7/8	205
80	15 3/8	11 3/4	8 1/4	231
90	15 3/8	12 1/4	8 5/8	253
100	15 3/8	12 3/4	9	264
120	17 1/8	13 1/8	9 1/2	304
150	17 1/8	14 1/4	9 7/8	370
200	19 3/8	15 3/4	11 1/4	526

Always slowly cool a crucible by leaving it in the furnace. Never leave any metal to cool in the crucible after pouring. It will swell upon re-heating and break the crucible.

Silicon carbide crucibles are much more durable and do not readily absorb moisture. They do slowly dissolve in iron. As with the clay graphite crucibles, do not leave any metal in the crucible to cool.

Ladles are used to take molten metal from a furnace to the waiting molds. They may be of the lip pouring type or the teapot type. Ladles may be made of coated or lined steel. They may be made from graphite and clay, silicon carbide or from vacuum formed ceramic fiber.

I often use a coated steel crucible to transfer 5 or 6 pounds of aluminum from the tilting furnace to small molds. Some furnaces have an open trough of molten metal called a "dip out well." Some ladles are made especially as dip out ladles.

You can make your own small ladles by welding a handle on a section of pipe or a sheet-metal bucket and lining it with refractory.

All refractories absorb moisture from the air, therefore all ladles should be preheated to drive off the moisture and bring the ladle up to the metal temperature. Small ladle heaters can be built from atmospheric type propane burners or fan type gas burners. Large ladle heaters may be oil fired.

Relined ladles must be thoroughly dried to prevent ladle explosions caused by pouring hot metal into an improperly dried ladle.

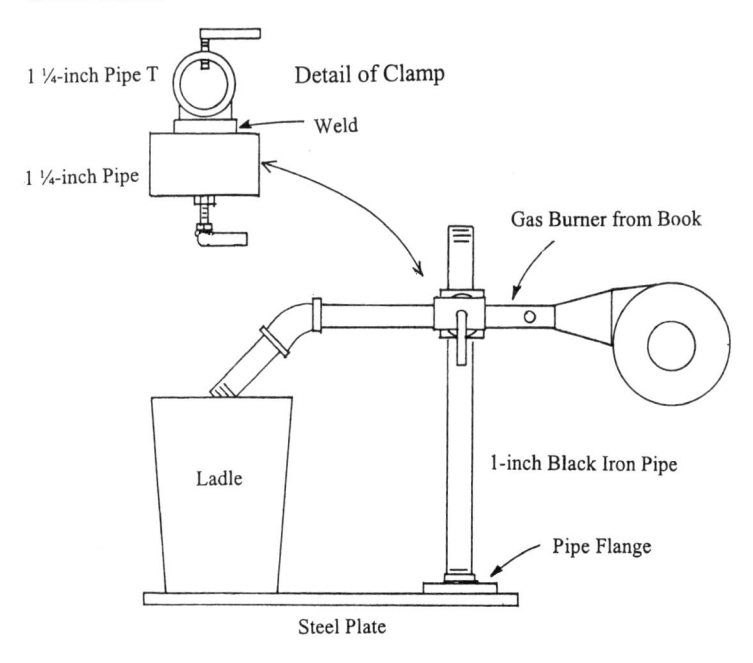

Ladle Heater

Cold Uncoverd Ladle   Preheated Ladle

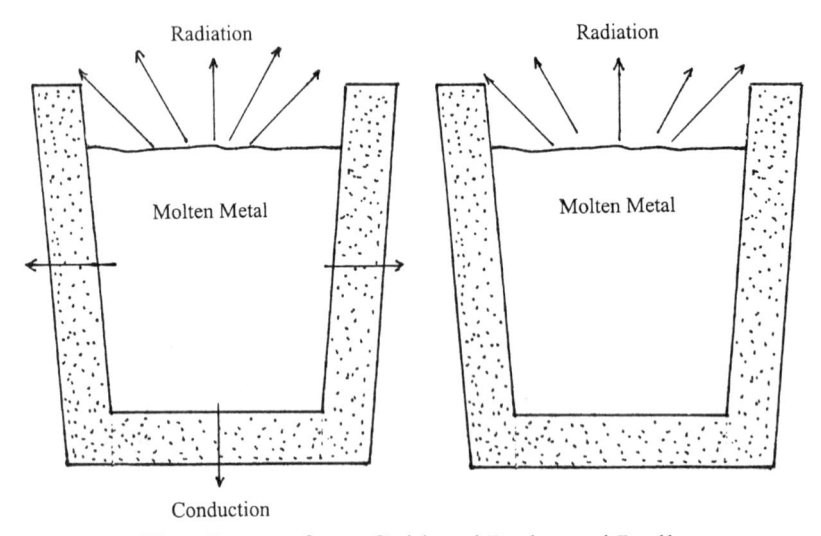

Heat Losses from Cold and Preheated Ladles

Heat losses in a ladle are from conduction through the walls and from radiation out the top. If the ladle is cold, the ladle absorbs heat from the metal until they are at the same temperature. If the ladle is preheated to the temperature of the metal, then the heat lost by conduction is small, however there is still considerable loss by radiation. Radiated heat increases rapidly with an increase in temperature. The increase is on the order of the fourth power as illustrated below:

$R = K(T_1^4 - T_2^4)$

R = radiated heat K = a constant

T_1 = metal temperature T_2 = temperature of atmosphere

Obviously, the temperature differences between the metal and the atmosphere are large. Pouring temperatures of some common metals are shown below:

Aluminum	1350°F	Yellow Brass	1950°F
Red Brass	2250°F	Cast Iron	2500°F

Covered Preheated Ladle

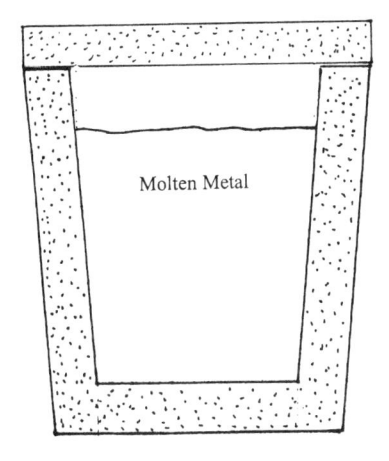

Molten Metal

Ceramic Ladle by Joy-Mark

You can see that the radiated heat from cast iron is considerably greater than that of aluminum. If an insulating cover is used, radiant heat loss is minimized.

One of my favorite ladles is vacuum-formed ceramic fiber over a wire reinforcing basket, made by Joymark. It is very light and it is extremely insulating. It can be used with all metals. The ladle is so insulating that it does not need preheating and may be dried by holding a gas weed burner over it for about 30 seconds before pouring molten metal into it. Aluminum does not stick to the ladle. Any aluminum coating or "skull" that remains in the ladle after cooling may be easily pulled out like a thick sheet of aluminum foil. Iron may be removed, but is more difficult to get out. Although the ladle is sold as a ladle liner, I often use a shank to hold it. Joymark does not make an insulating cover for the ladle (yet). When melting iron in the cupola, I usually cover my ladle with another ladle between taps to minimize the radiant heat losses.

Calculating ladle capacity:

Ladle capacity may be easily calculated by multiplying the volume times the density of the metal as shown below.

$$V = .196 \times (D_T + D_B)^2 \times H$$

V = volume
D_T = inside diameter at the top of metal
D_B = inside diameter of bottom
H = height of metal in ladle

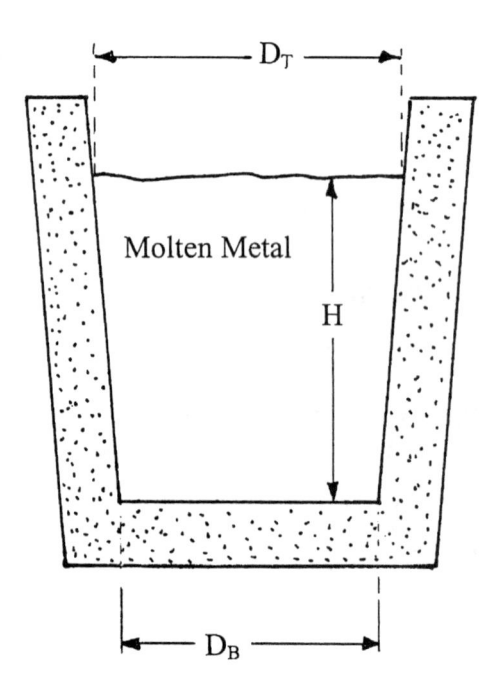

Ladle Capacity in pounds = Volume x Density of Metal

Densities of Molten Metals

Metal	lb/cubic inch
Aluminum	0.079
Brass	0.243
Copper	0.288
Cast Iron	0.238
Lead	0.379
Steel	0.250
Zinc	0.230
Tin	0.250

Calculate the capacity of a ladle for aluminum, brass and cast iron. The ladle has a bottom inside diameter of 6.5 inches, is filled 8 inches deep with molten metal and has an inside diameter of 8 inches at the top of the molten metal:

D_T = 8 inches D_B = 6.5 inches   H = 8 inches

$V = .196 \times (8 + 6.5)^2 \times 8 = 330$

Volume = 330 cubic inches

Capacity in pounds = Volume x Density

For aluminum:

Capacity = (330 inch^3 x .079 pounds / inch^3) = 26 pounds

Capacity of Brass = 80 pounds

Capacity of Iron = 78.5 pounds

FURNACES FOR THE SMALL FOUNDRY:

Furnaces may easily be built for little cost. Homemade furnaces can cost as little as twenty dollars for a small unit that will melt a quart or about 5½ pounds of aluminum. Larger units that melt a few hundred pounds of iron or one hundred pounds of aluminum per hour may be built for about two hundred dollars.

Furnaces may be fueled with grocery store charcoal, coke, propane, diesel fuel or used motor oil. Small electric furnaces may also be built for little cost.

Fuel fired furnaces are either direct-fired or indirect fired. In a direct fired furnace the products of combustion come into contact with the metal charge as in the case of the reverbertory furnace and the cupola furnace. When the products of combustion do not come into contact with the metal, the furnaces are called indirect-fired. A crucible furnace is an example of an indirect-fired furnace. Different types of furnaces influence the melting losses and the chemical composition of the metal. Direct-fired furnaces are more efficient than indirect furnaces; however melting losses are greater.

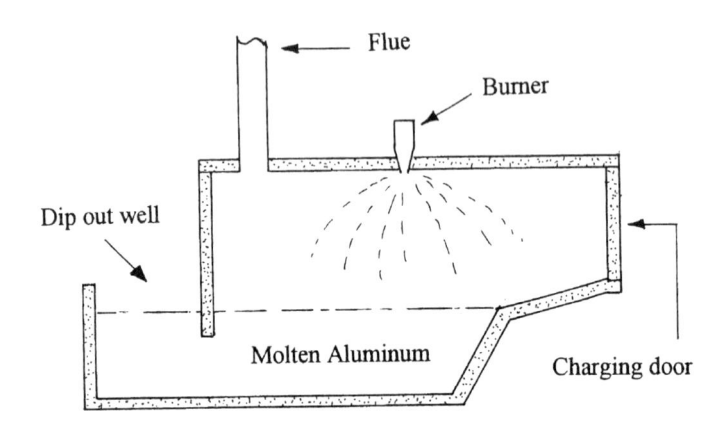

Reverberatory Furnace

Crucible furnaces are not as efficient as reverberatory furnaces because the heat must be transferred through the walls of a crucible to the metal charge. Crucible furnaces come in three types, the lift out crucible type, where the

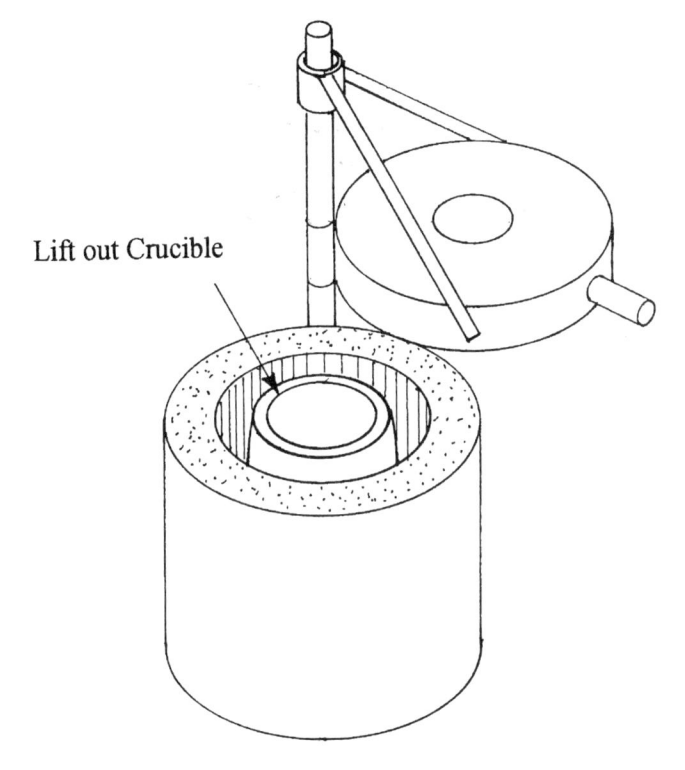

Lift out Crucible

Lift-out Crucible Furnace

crucible is used as a ladle, the stationary crucible type where the aluminum is dipped out with a ladle and the tilting type where the metal is poured directly from the furnace. Crucible furnaces generally have between 3 to 5 inches of refractory that is rammed or poured inside a steel shell.

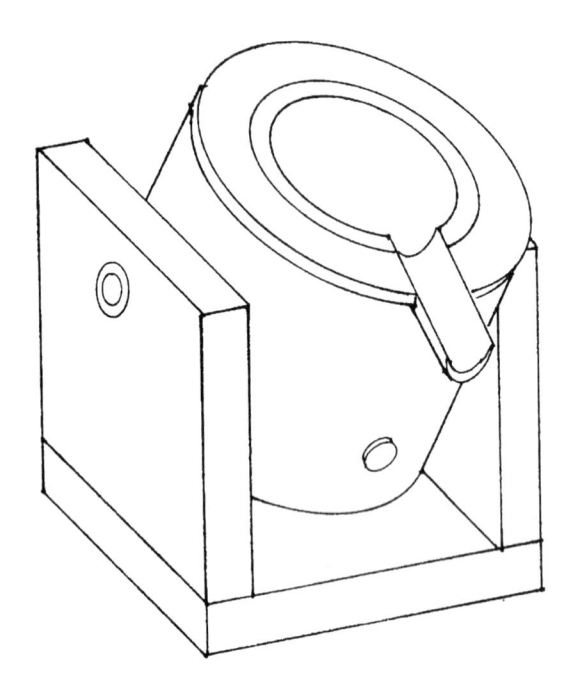

Tilting Crucible furnace

Cupolas or "Shaft Furnaces" have been used to melt iron for hundreds of years in the west, and the Chinese have used cupola type devices for thousands of years. Small cupolas were common in the USA until the end of WWII when larger foundries bought up many small foundries or they could not compete with the larger, more efficient operations. Thus the decline of the small job shop with a cupola.

Cupola furnaces are simple to operate; however operation is an exact science. The basic idea is you dump weighed charges of iron and coke in the top and tap molten iron out the bottom. A blower supplies air to the fire in the cupola. The air or "blast" enters the sides of cupola through holes called "tuyeres" (tweers). The blast keeps the coke

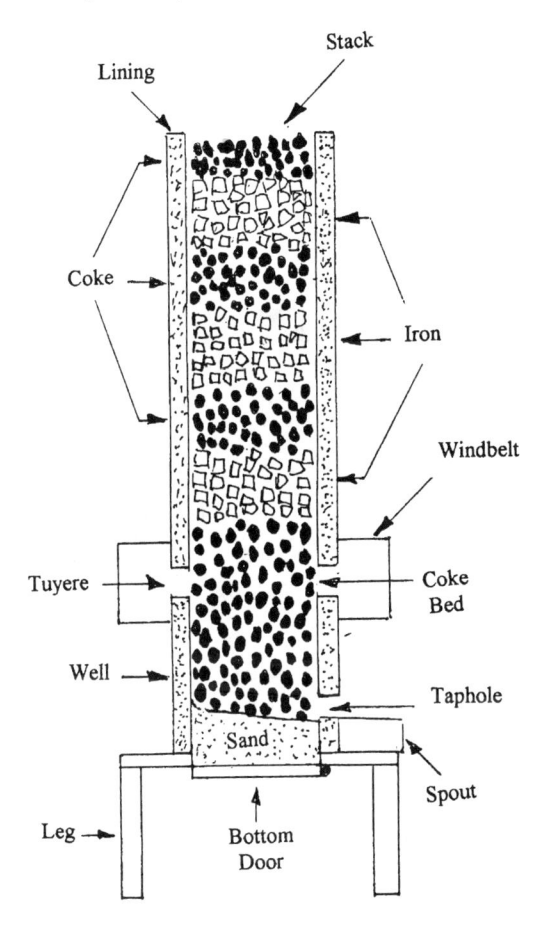

Cutaway of Cupola Furnace

bed at white heat. The carefully weighed charges are calculated so that as the coke bed burns away, it is constantly being replenished with new coke. Thus the height of white hot coke bed in the cupola remains constant. A properly maintained coke bed ensures iron taps at 2750 degrees or more. Bronze may also be melted in a cupola furnace.

Cupolas are fast melters and are not in the same league as gas furnaces. A gas furnace may take an hour or more to melt a ladle of iron. Cupolas will melt a ladle every 6 to 12 minutes. Because cupolas are such fast melters, you must have all of your molds ready ahead of time. Usually I will make molds for a day or two then I will fire up the cupola and fill them all in less than an hour.

Tapping a Ladle of Iron from a 10-inch Cupola

BEGINNER'S CHARCOAL FURNACE:

A "Gingery" type charcoal furnace is a good project for a beginner. It melts a quart of aluminum in about 20 minutes over a bed of charcoal. You can easily build the furnace in few hours.

The furnace is made from a 5-gallon metal bucket lined with refractory. You can use a homemade mix or buy commercial castable refractory. The homemade mix is cheap and easy. The castable refractory will last longer. I usually get 30 to 50 heats from the homemade mix.

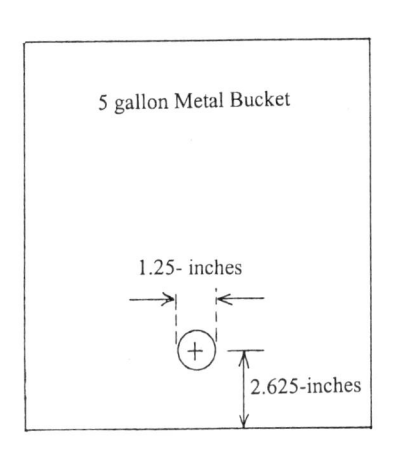

Start by drilling a 1¼-inch hole in the bucket 2.625-inches up from the bottom. A hole saw works well or you may scribe a circle, punch and drill several small holes around the circle and then knock out the center section with a chisel.

```
5 gallon Metal Bucket

1.25- inches
  →| |←
    (+)
        2.625-inches
```

THE LINING:

The refractory mixture is made from one 50-pound bag of coarse sand, one 50-pound bag of fine sand and one 50-pound bag of fireclay. Fireclay may be purchased at masonry supply stores and at building supply stores who cater to masons building fireplaces. Fireclay cost approximately $8 per 50 pounds.

Pour half a bag of each sand into a wheelbarrow and add a little water. Turn it over until it is well mixed then slowly add ½ bag of fireclay while turning it over for quite awhile. It must be well mixed, so be patient. When it is

Crucible – Ladle

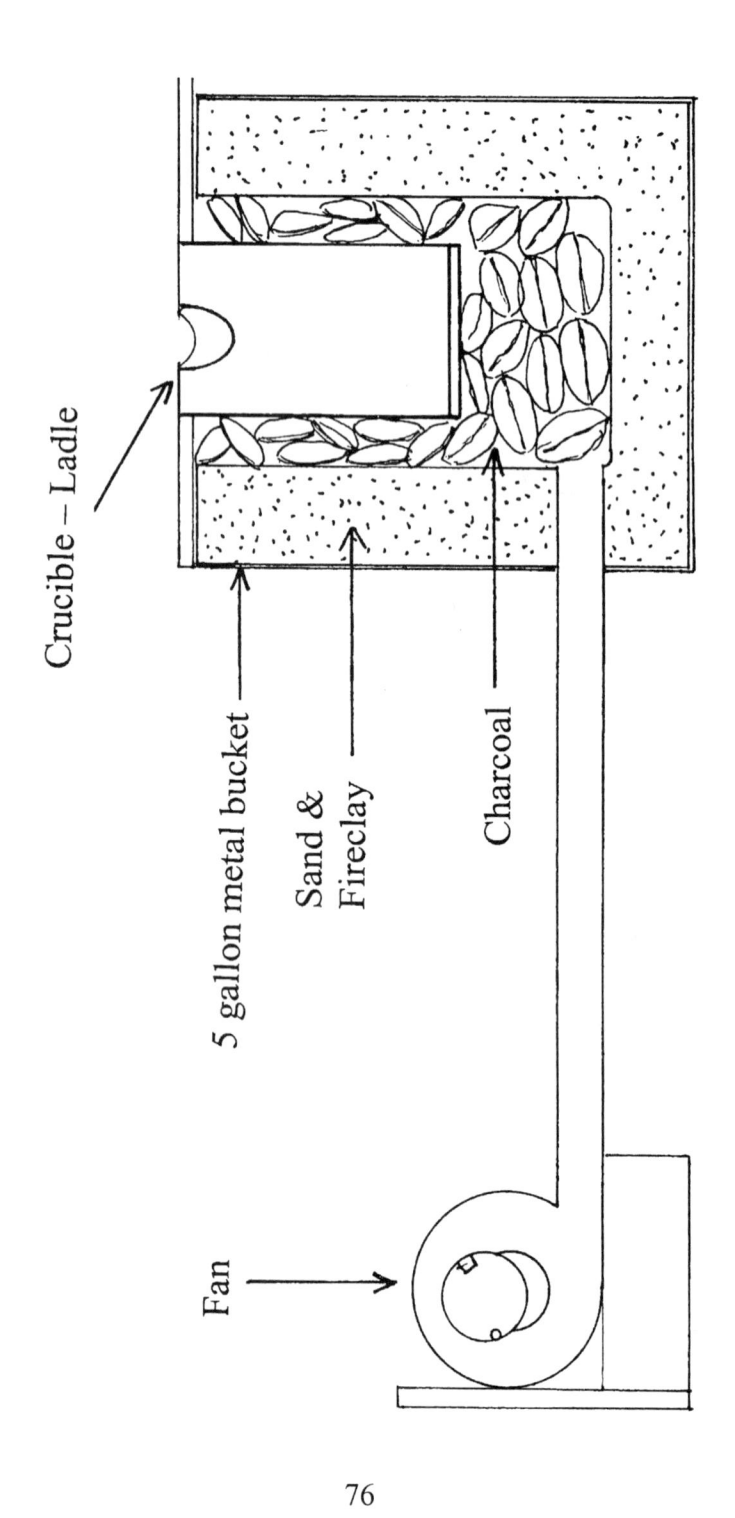

5 gallon metal bucket

Sand & Fireclay

Charcoal

Fan

well mixed, start adding the rest of the water by misting with a garden hose. Don't let the water puddle in the mixture or the clay will separate from the sand. Clay takes up water slowly. This process takes a while, so be prepared to do some shoveling. When the mixture has the consistency of stiff mortar, it is about right. Don't let the mixture become too wet, it will not cure well and will slough off when fired. When you are satisfied with the consistency, cover it with plastic and take a break for a few hours or overnight.

MAKING THE INSIDE FORM:

Make the form by wrapping 24-gauge sheet metal around two 7-inch plywood discs. These can be cut with a jig-saw, however turning them on a lathe is better. A 1-inch deep slot is cut in one edge of each disk. The top and bottom disks have a stout screw put in one side. When removing the disk from the furnace you will grab this screw with a pair of pliers and pull it up. Cut a piece of sheet metal 11-inches high. The length of the piece is the circumference of the disc plus 1-inch. For a 7-inch disc, this would be 23 inches long. Make a right angle bend 1-inch deep along the 10-inch side. This will hook into the saw slot of each disk. Tape the sheet metal in place with duct tape. If the metal is springy and difficult to work, roll it up to a diameter less than 10 inches and try taping it again.

Refractory Form

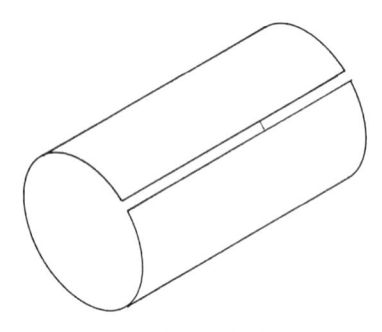

Get a 2 or 3-foot section of galvanized top rail used for chain link fence. It is thin pipe and easily cut. Saw off two 3-inch sections. Clamp a section in a vise and saw a slot down one side so that the pipe may be sprung open. Repeat the process on another 3-inch section of pipe. Saw off a 4-inch section of pipe for the tuyere or air input to the furnace. The other sections form small clamps to attach a blower.

LINING THE FURNACE:

Put a layer of refractory in the furnace and with a section of 2 x 4, ram it tightly against the bottom. Continue until the bottom is 2-inches deep and reaches the bottom of the 1¼-inch hole. The furnace is lined by ramming refractory around the form. The form is located relative to the shell by using four wood strips at 90 degrees to each other. The strips are 2¼ inches wide and 18 inches long. Insert the 4-inch section of pipe cut earlier into the hole and butt it up against the form. Ram the refractory around the form and raise the wood strips as each layer of refractory is laid.

After ramming the lining, peel the tape off the top disk, grab the screw with pliers and remove the disk. Remove the bottom disk in a similar manner. Collapse the sheet metal and remove it. Don't be concerned if the form has collapsed a little out of round while ramming the lining. Patch any voids in the lining and make a smooth fillet between the sides and the bottom. Using a piece of welding rod, poke holes every inch or two around the lining. This is to vent steam from the lining during the curing process.

Screw

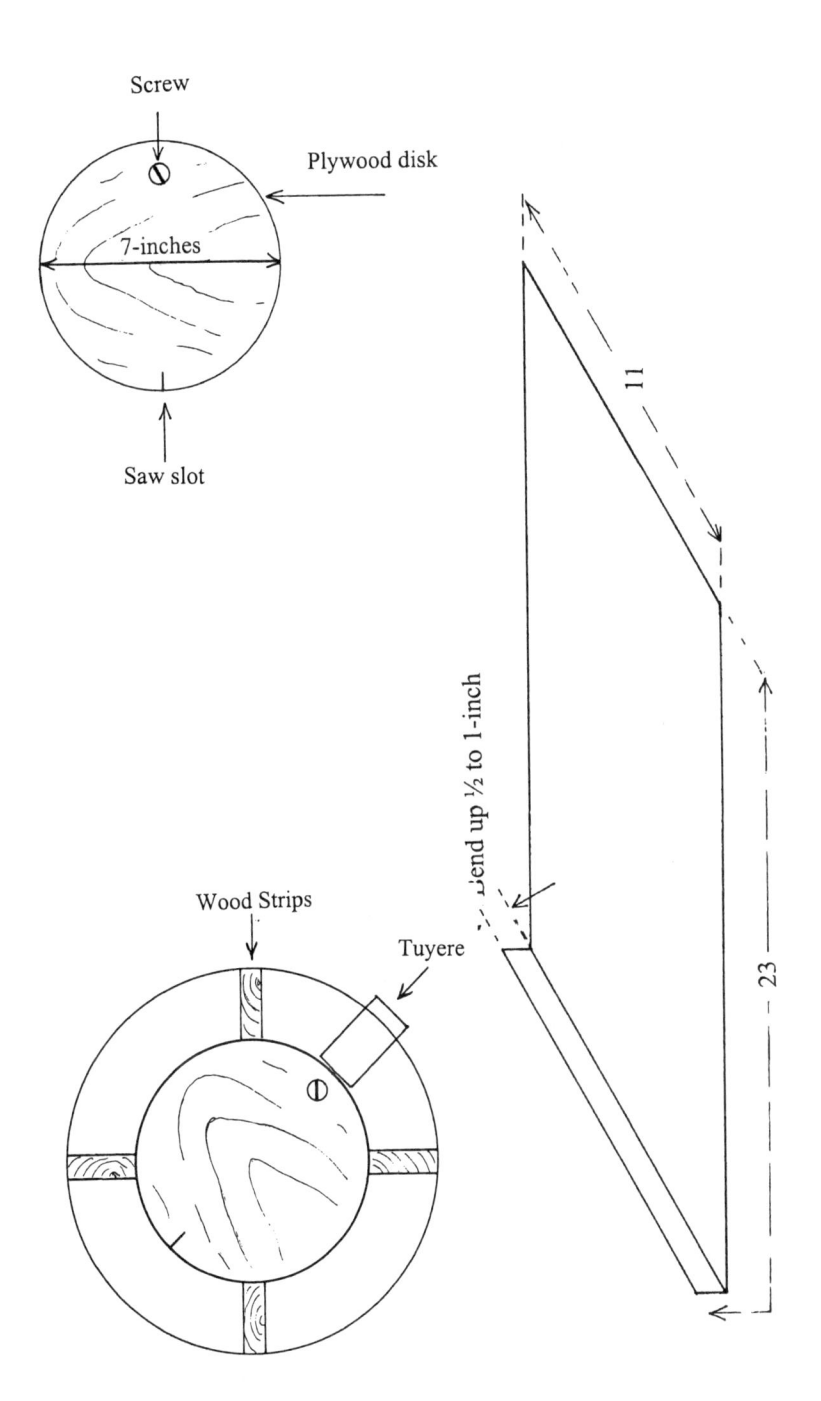

Plywood disk

7-inches

Saw slot

Wood Strips

Tuyere

bend up ½ to 1-inch

11

23

CURING THE LINING:

Soak several wads of newspaper in starter fluid and throw them into the furnace. Soak enough charcoal in starter fluid to make three layers in the bottom. Light the furnace by placing a lit propane torch through the airhole. Turn on the blower to help get the fire started. When the charcoal is burning well, turn off the blower and add more charcoal. This could take a few hours, but you will want to have the whole furnace full of burning charcoal. After two hours of burning, turn on the blower and adjust it for low airflow. After a half an hour turn it on full blast for 15 minutes to half and hour then turn it off and let the furnace sit over night to cool.

Firing the lining is a slow process. If you get the lining too hot too fast, the outer surface of the lining will vitrify but the inside will still be wet. When you turn the heat up, steam will build up in the lining causing it to crack and slough off. Be patient and take your time.

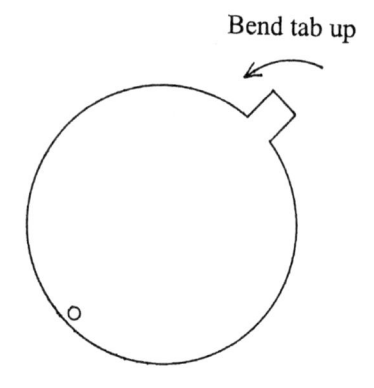

You need a small blower to operate the furnace. A blow-drier works for starters but is a little noisy. You may purchase a small blower for about $10 at Wholesale Tool supply or an equivalent vendor. Make a shutter for the blower so you can turn down the air blast.

Sheet-metal Shutter

MAKING THE CRUCIBLE:

The crucible may be made from 7¼-inch length of 4-inch diameter pipe with a plate welded onto the bottom.

Cut the bottom from 1/4-inch plate and grind a bevel on the edge as shown in the diagram. The bevel ensures good penetration on the welded joint.

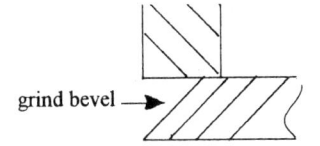

grind bevel →

Good

No Good

Be sure that the base of the pipe and the plate are clean and free of rust, dirt and oil, then weld the baseplate to the pipe. Carefully inspect your weld for defects. Fill the crucible with water and check for leaks. If any defects or leaks are found, grind out the bad weld and re-weld it. Do not let the bottom stick out past the edge of the pipe diameter or the crucible will hang up when trying to get it out of the furnace. Heat up a spot on the top edge of the pipe with a torch and form a pouring spout.

Weld a 4-inch section of 3/8-inch rod to a 2-foot section of rod to form a "T." Get a 5-inch section of "closet rod" from a building supply store. It is 1½ diameter dowel used for hanging clothes. Drill a 3/8-inch diameter hole through the center of the rod to make a sliding handle. Slide the dowel over the 2-foot section of the "T" and weld it to the crucible to form a pouring handle. Cut another 4 or 5-inch section of 3/8-inch rod and weld it opposite the handle to form an ear on the top of the crucible. The crucible will hang in the top of the furnace while melting. Keep the wooden handle out at the cool end of the handle when the furnace is operating. Slide it up to the crucible for pouring. For most castings, the iron pot will work fine. Where the maximum properties of the aluminum are needed, the iron dissolved from the pot will make the aluminum more brittle. You will have to coat the inside of the pot with ladle wash to prevent iron pickup in the aluminum. A home brew ladle wash is described in the melting section or it may be purchased from Mifco and Budget Casting Supply among others. Purchased ladle wash is usually a solution of boron nitride.

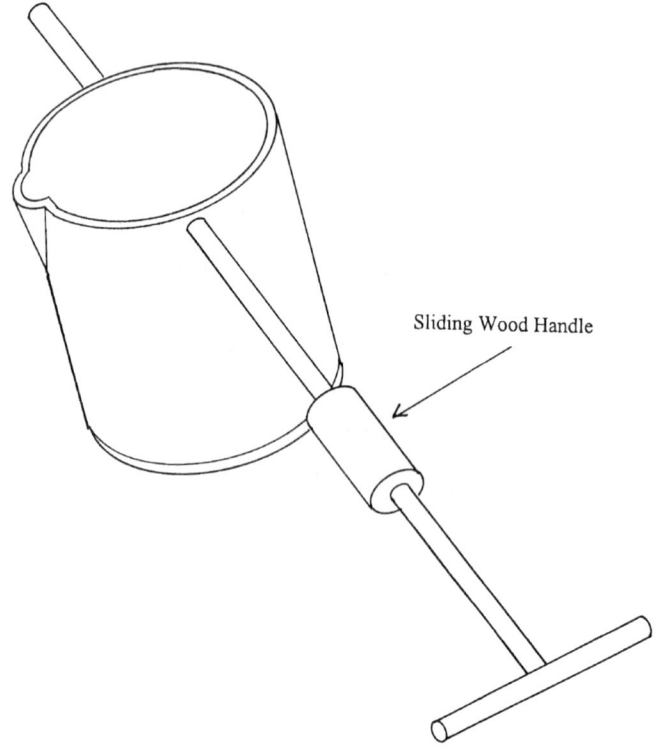

Sliding Wood Handle

The furnace seems to melt best when the charcoal in the bottom of the furnace is full sized briquettes and the rest of the charcoal is smaller. The charcoal splits cleanly if you stand it up on edge and hit the top with a hammer.

Hit Here Splits Cleanly

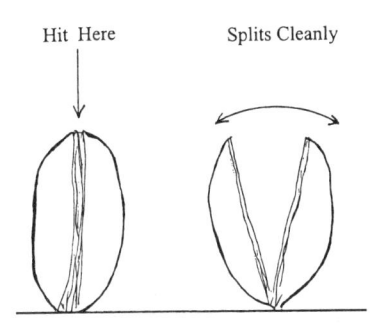

Start with a low blast and slowly increase it to get the best melting. Be careful not to use too much blast and blow all the heat out of the furnace.

Lifting a Crucible from the Charcoal Furnace Using Home Made Tongs

STEPPING UP TO A GAS FIRED FURNACE:

After you have tried the charcoal furnace and decided that metal casting is for you, you may want to build a gas fired furnace. The gas burner eliminates the smoke and ash associated with starting and operating a charcoal furnace. (Note: the charcoal furnace only smokes when being lit, similar to a grill.) The gas burner allows you to operate for longer periods of time because you do not have to stop and remove the charcoal ashes from the furnace.

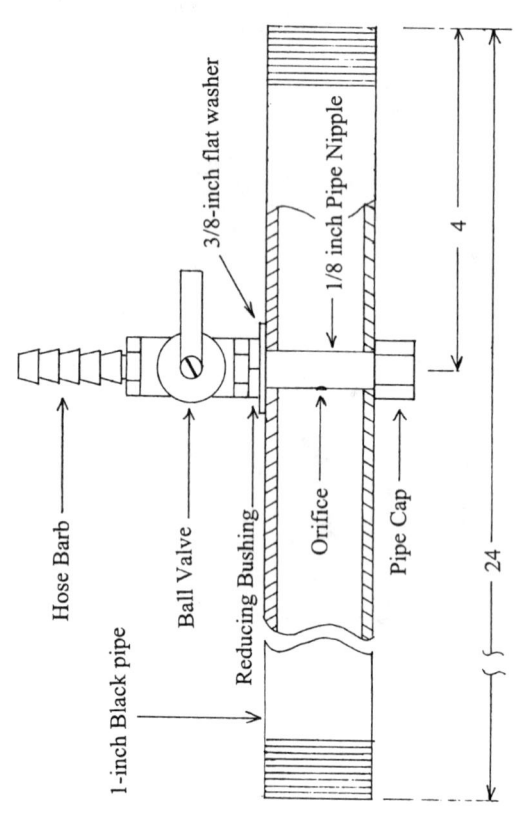

The gas burner, requiring only a drill and a few hand tools, is simple to make. It fits directly into the burner hole of the charcoal furnace so it may be easily substituted in your existing furnace.

The burner is made from common plumbing pipe and fittings. The sizes of the fittings are selected so that the burner may be assembled without welding. Start with a 2-inch long 1/8-inch pipe nipple and pipe cap. These may be found at true hardware stores such as "Ace Hardware." The Home Depot types of stores only stock fast selling items and are not likely to have the smaller sizes of pipe.

Wrap one end of the pipe nipple with Teflon tape and securely screw on the pipe cap. Hold the assembly up to one end of the 1-inch pipe and using a sharpie pen, mark the center of the pipe on the nipple. This is the location of the gas jet hole.

Arrow Points Toward Orifice

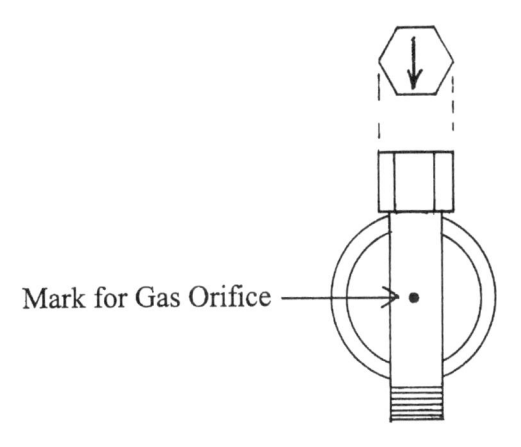

Mark for Gas Orifice

*Note: natural gas requires a much larger hole, 1/4 -inch or 17/64-inch.

Center punch the hole and clamp or place the nipple in a vise to drill the gas jet. Using a 1/16-inch* jobbers drill, drill the hole through one side of the nipple. (An inexpensive set of jobber drills may be purchased from Enco, or similar tool dealers.) Select a larger bit, approximately 1/8-inch, and make a slight cut to flare the hole previously drilled. Remove the nipple from the vise and mark an arrow on the pipe cap that points in the direction of the hole. Later, this arrow will help you align the orifice in the burner tube.

Flairing Cut on Gas Orifice

Wrap a ¼ to 1/8-inch reducing bushing with Teflon tape and securely screw it into a ¼-inch ball valve. Wrap a ¼ inch hose barb with Teflon tape and screw it into the ball valve. Set this assembly aside while you make the burner tube.

Make the burner tube from a 2-foot section of 1-inch black pipe. Drill a 13/32-inch diameter hole completely through the pipe 4-inches from one end. Insert the gas jet assembled earlier through the hole and wrap Teflon tape over the exposed threads. Place a 3/8-inch flat washer over the nipple and screw on the reducing bushing and valve previously assembled. Using the arrow previously drawn, align the orifice to point down the center of the long side of the 1-inch pipe. Tighten the bushing-valve assembly down while holding the orifice in alignment. Later you may attach the gas hose to the hose barb with a hose clamp or you may substitute a fitting for left-hand threads used on gas hoses.

For higher temperatures and more efficient melting, tangential placement of the burner is required. This forms a "cyclone" combustion chamber where the flame swirls around the crucible. The furnace also requires a lid to help retain the heat. These changes are detailed below.

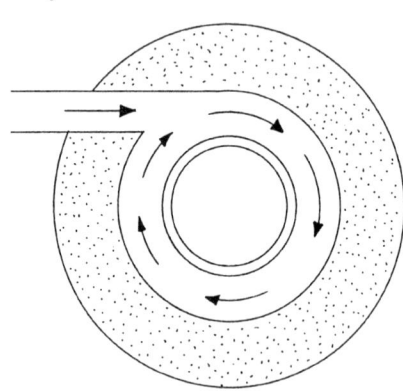

Because you do not need extra space for charcoal ashes in the furnace, you can raise the burner entrance up to 4 inches from the bottom of the bucket. Make a paper template as described in the "sheet metal forming" chapter

Cyclone Combustion Chamber

Exhaust Port
1 ½-inch iron pipe

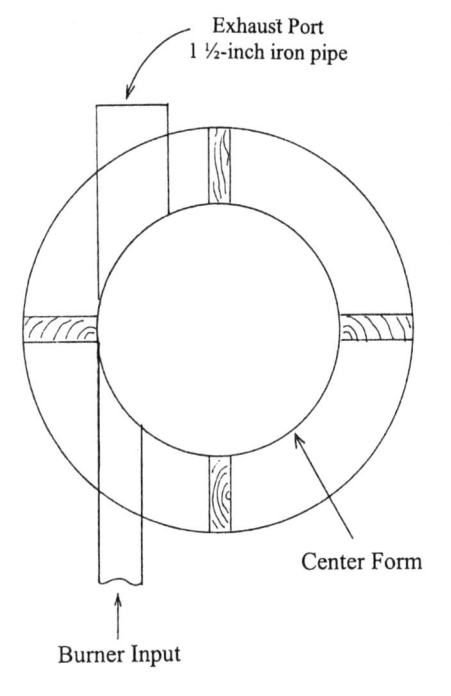

Center Form

Burner Input

to mark the cutout for the intersection of the burner and the furnace body. Align the template on the furnace body and mist the area with spray paint to mark the cutout. Using a sharpie pen, draw a good outline, and check it with the template. Prick-punch the perimeter of the cut out and drill it with a small drill bit. Using a chisel, knock out the center and then clean up the hole with a file, if needed.

Add an additional exhaust port at the top of the furnace body. This allows the combustion gasses to escape without having to seep around the top of the crucible in order to get to the exhaust port in the lid.

If you are planning to use the burner only for this furnace and not for a ladle heater, you may want to make a template for the inside diameter of the furnace and trim the burner tube as seen in the drawing. I use the burner in several different applications, so I did not cut the end to match the inside diameter. In this case, cut and grind another small section of pipe or fence top rail described earlier to match the inside diameter. This is used to form a void in the refractory similar to that used in the charcoal furnace. Ram and vent the refractory as was done in the charcoal furnace. The refractory is easily smoothed around the pipe with a wet finger.

Make the lid from a strip of sheet metal that is 2½ - inches wide and an inch longer than the circumference of the pail. I am using 24 gage galvanized sheet steel, however the zinc coating oxidizes at furnace temperatures. Unplated steel is better for furnace construction. Measure and drill 8 equally spaced 1/8-inch holes in the strip. To give it a circular bend, the strip may be formed around a section of pipe or wood that is rounded and clamped in a vise.

Using a vise-grip, clamp the ends of the strip together. Punch and drill the ends to for sheet-metal screws or rivets and fasten the strip together. Using re-bar tie wire or similar, thread it through the holes as shown in the drawing. After the last hole, twist the ends of the wire together. Cut and bend the 11.5 x 2-inch strip of 16 gage and bolt it to

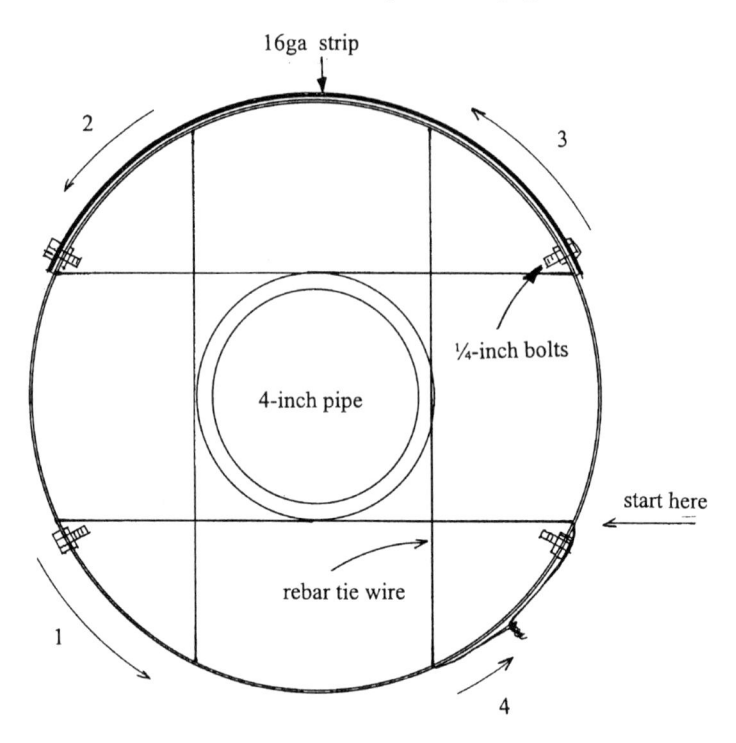

Furnace Lid

one side of the lid frame. Bend and drill two strips of sheet metal to attach the chain to the lid. When bending, insert a strip of scrap metal between the legs of the strip to make clearance for attaching it to the lid. Insert a section of 4-inch pipe to form the center hole. Adjust the wire as needed so that the hoop remains round.

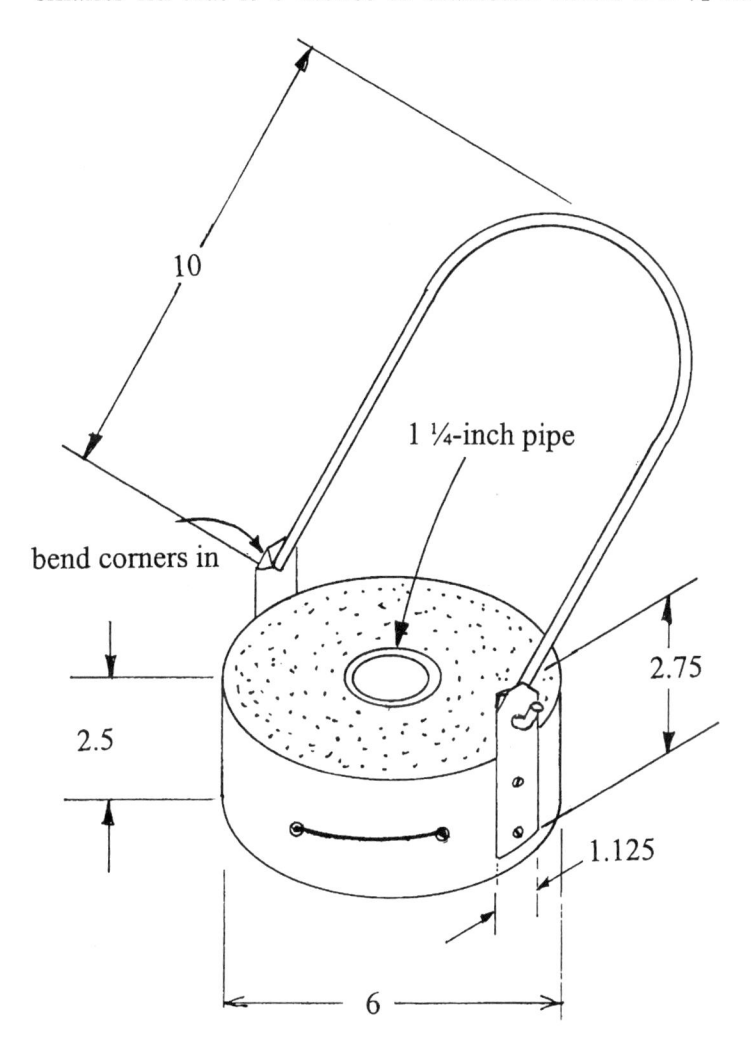

Using a 20-inch strip of sheet metal, make a second smaller lid that is 6 inches in diameter. Insert a 2 ½-inch length of 1 ¼-inch black pipe. Leave it in place when firing the lid. Bend a handle from stiff wire, like you would find in a paper sign. Attach it to the furnace lid using two ears as seen in the drawing. Bend the corners of the ears in so that the wire handle can not fall against the side of the furnace and get hot from escaping gasses. When melting, you want to keep it in the 45-degree position so that the gasses from the open port in the lid will not make the handle too hot to touch.

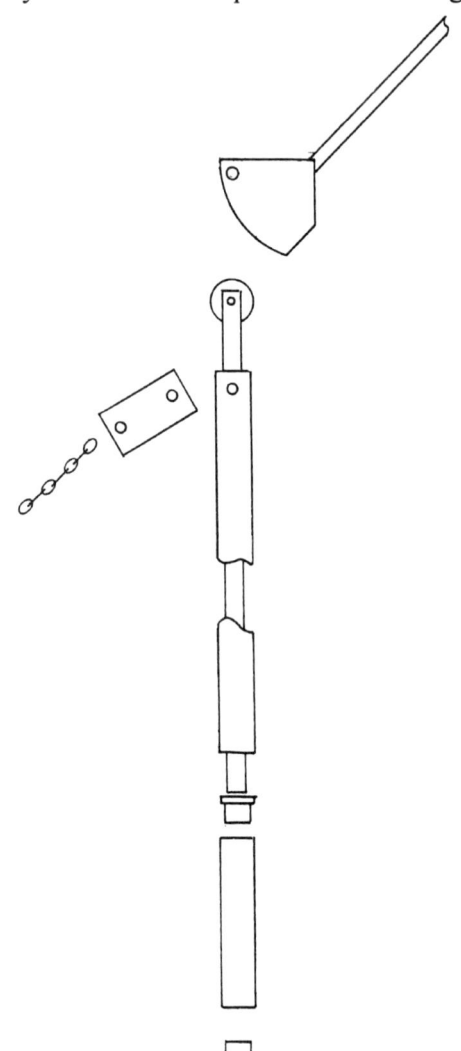

The lids may be lined with castable refractory or with the sand a fireclay mixture used for the charcoal furnace. Set the lid on a section of stiff cardboard or thin plywood to ram the refractory. It you are using the sand and fireclay mixture, it should be rammed with a wood block and a hammer. After the larger lid has been rammed, spin the 4-inch iron pipe to loosen it. It may then be pulled out leaving a smooth hole in the lid. Carefully move the lid to the furnace for slow curing over a low charcoal fire. After it has dried over a low fire for a few hours, turn up the air for curing at high temperature for an hour. Finish the furnace and lid by turning on the gas burner to bring the inside up to a red heat. Turn off the gas and remove the burner. Place the smaller lid over the hole and allow the furnace to cool slowly over a few hours.

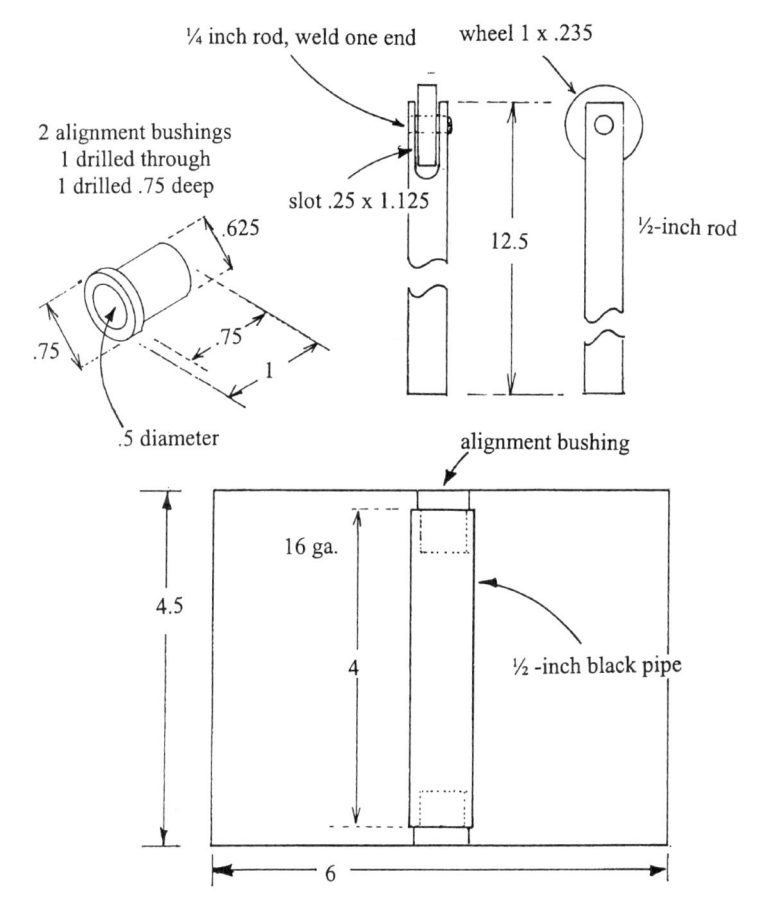

¼ inch rod, weld one end   wheel 1 x .235

2 alignment bushings
1 drilled through
1 drilled .75 deep

slot .25 x 1.125

.625

12.5

½-inch rod

.75

.75

1

.5 diameter

alignment bushing

16 ga.

4.5

4

½ -inch black pipe

6

Building the lifting device: The lid is raised using a lever operated cam over a roller as seen above.

Make two alignment bushings from ¾-inch diameter steel rod. The top bushing is drilled through using a ½-inch drill while the bottom one is drilled only ¾-inch deep. Using a 4-inch section of ½ inch black pipe, insert the bushings and weld them in place. Using 16 gage material, heat and form a 4 ½ by 6 inch plate to match the outside of the furnace. Weld the assembly together as seen in the lower part of the drawing.

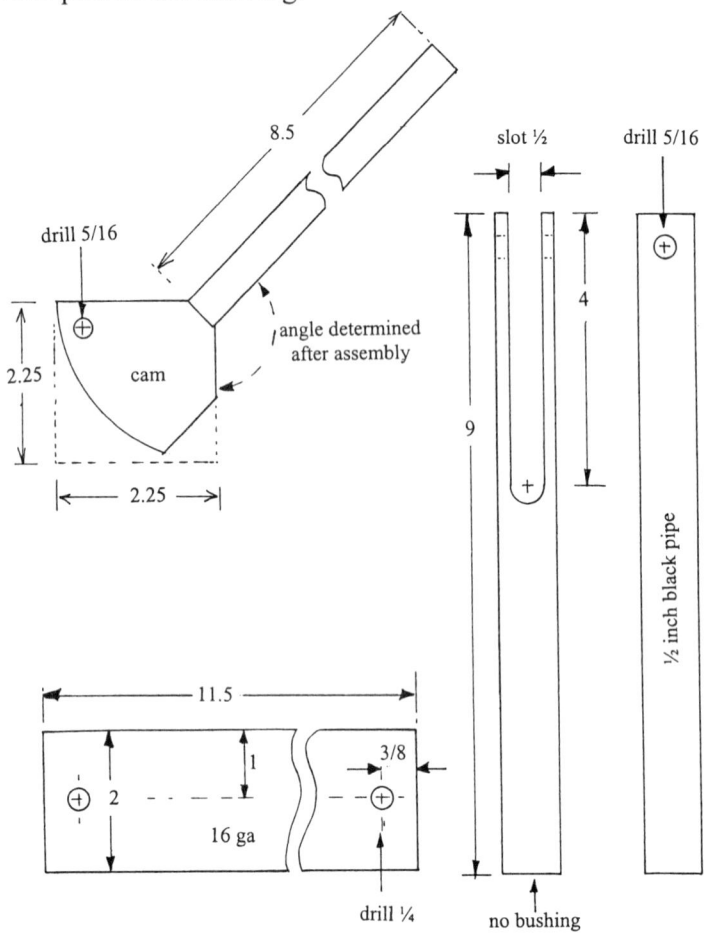

Make the roller rod from a 12 ½-inch length of ½ inch rod. Drill a ¼-inch hole 3/16 (.1875)-inch down from the end. The slot may be cut by drilling a ¼-inch diameter hole in the rod and sawing out the excess material. Clean up the cut with a file. It is easier to mill the slot with a ¼ inch end mill. Cut and file to thickness, a wheel from a 1-inch diameter steel or brass rod. Punch and drill the center hole ¼-inch diameter. This entire operation is quickly and easily performed on a lathe. Assemble the wheel and rod using a section of ¼-inch steel rod. Weld one end to hold the assembly together.

The cam is cut from a section of ½ inch thick steel plate. The thickness can be 3/8 of an inch, if that material is available. This furnace was built from my scrap pile so you may adjust the materials as necessary. The radius is rough sawed and finished on a grinder. A flat is cut on the bottom to give the cam a stable "lifted position." The handle may be welded in position parallel to the flat on the cam or you may wait until the furnace is assembled to find the most comfortable position for yourself.

Make the cam holder from a 9-inch section of ½-inch black iron pipe. Drill a 5/16-inch diameter hole centered approximately ½-inch down from the top of the pipe. Roll the pipe 90-degrees and cut the ½ inch wide slot 4-inches long. The width of the slot depends upon the thickness of the cam. If you are using 3/8-inch stock, the slot should be 3/8-inch wide. To prevent binding when the furnace is hot, use no bushings in the cam holder pipe.

Insert the roller rod into the cam holder (½-inch diameter pipe with slot). Bolt the cam and two sheet-metal strips to hold the cover chain to the cam holder using a 5/16-inch diameter bolt.

Attaching the lid: Square the bushing assembly to the furnace body and weld into position. The thin metal of the 5-gallon bucket burns through easily. Use .023 wire and low heat. Set the large furnace lid into position with the 16-gage strip facing the cam holder tube. Insert the cam and holder assembly into the bushings. Adjust the lid as necessary and weld the cam holder to the strip Bend the

Notch edge of furnace for crucible. strips out at an appropriate angle and attach the chain by looping it through the holes and securing the ends using tie ware. The fine chain as is commonly used to hang fluorescent lights works well for the furnace lid.

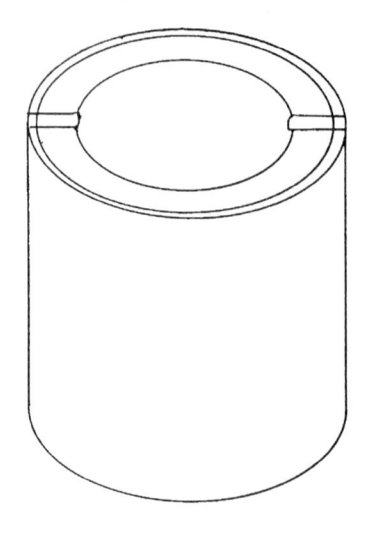

Depending upon how well your lid fits, the handles will get hot. They may become hot enough to be difficult to hold while wearing gloves. If you are going to melt with the welded crucible used in the charcoal furnace, you will want to cut a relief in the uncured refractory so that the lid sits down flush on the top of the furnace.

Charging Hopper

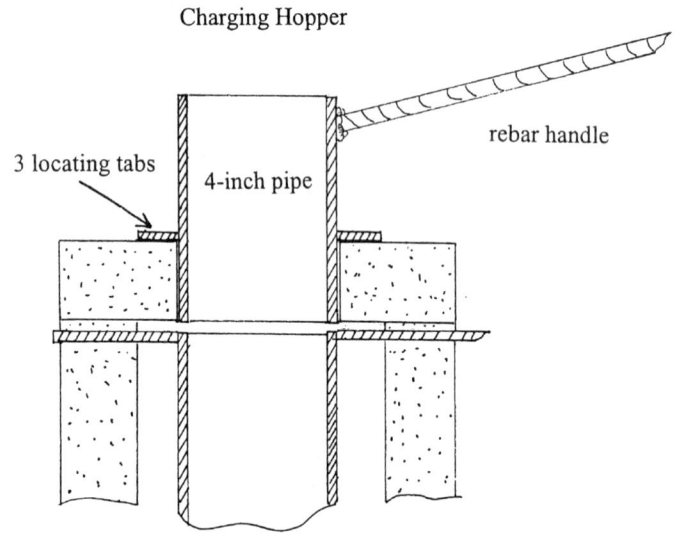

A charging hopper speeds melting by preheating the charge. It also dries and burns off residual oils that may gas the melt. The proper way to use a charging hopper is to allow the material to preheat and then push the hot material into the melt. To prevent oxidation of the metal, do not melt material in the charging hopper and let it run down into the crucible.

If you are melting higher temperature metals and need to set the crucible on a plinth, make three or four by ramming refractory into a form about 3.5 to 4 inches in diameter and 1.5 to 2 inches deep. Cut two groves approximately ½-inch deep as shown in the drawing. Fire each one in the furnace. Place a section of cardboard on a new plinth to prevent the crucible from sticking to it.

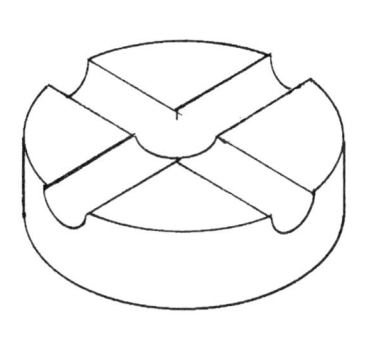

GAS TANKS AND REGULATORS:

This small furnace has been designed using the tables and formulas found in *"Build an Oil Fired Tilting Furnace."* It is designed to melt 6 pounds of aluminum every 20 minutes. The burner generates 93,000 Btu per hour with a propane input pressure of 5 psi. The orifice flows 36 cubic feet per hour at this pressure. The maximum flow at 10 psi is 130,000 Btu. Small regulators are NOT likely to work. You may use an adjustable propane regulator or, I prefer to use an acetylene regulator found on gas welding equipment. They have fine adjustment and work very well. Prices vary considerably, so shop around. Harbor Freight sells a regulator for approximately $39. Propane is a solvent to the rubber found in gas welding hoses. The hoses deteriorate more quickly so inspect them frequently or use propane rated hoses.

A common complaint when melting with gas is "there is bad gas at the bottom of the tank." The gas is not bad, however the tank is probably undersized for the burner. Tanks have a specific "vaporization capacity" depending on the temperature and the amount of propane in the tank. Using the vaporization charts and diagrams found in *"Build an Oil Fired Tilting Furnace"* a 100-pound cylinder is the minimum size recommended to properly operate this furnace. These cylinders may be purchased or rented from your local gas supplier for little cost.

Completed Furnace

LIGHTING AND OPERATING THE FURNACE:

Set the furnace up outside away from anything that may catch on fire. Attach the blower and burner to the furnace using the split pipe couplings described in the charcoal furnace section. Attach the fuel line to the furnace. Make sure that the gas tank is a safe distance away from the work area. The gas line must be out of the way and not likely to be tripped over or damaged by spilled metal. Adjust the gas regulator for 3 psi.

Make a lighting torch by bending the end of a straightened coat hanger over a folded strip of paper towel. Dip the end of the torch in kerosene, lighter fluid or spray it with WD-40. Light it and push it through the hole in the lid of the furnace. Turn on the blower with the air shutter almost completely closed. Open the ball valve on the gas supply. The furnace should light in 2 to 3 seconds. Open the air shutter and adjust the gas up to 5 psi. Fine-tune the air for the maximum combustion noise. Remove the lighting torch after the furnace has been operating for 10 to 20 seconds. If the furnace does not light in 5 seconds, turn off the gas, and review all the connections and settings.

FORMING SHEET METAL FOR FURNACES:

There are two types of sheet metal forms that are used in small furnace work. They are *transition pieces* used to join two pipes or ducts of different sizes and the *intersection of right cylinders* used to join pipes tangent to a furnace body. Transition pieces may join a square or rectangular duct to a round duct or they may join two round ducts of different size.

Sheet metal development is not difficult and is a very useful skill to have. It takes much longer to describe the process than it takes to actually do it. You do not have to have a sheet metal brake to make the forms, however compound aviation style tin snips are good for this type of work. When cutting with aviation snips, do not completely close the snips at the end of a cut. This bends the sheet metal leaving a bump that must be dressed later. Rough edges may be pounded flat and smoothed with a file.

Make small angle bends by clamping the sheet metal to a workbench top under a strip of angle iron and bending with a wooden mallet, soft face hammer, or a hammer and block of wood. Make large angle bends in small steps, no more than 30 degrees at a time. Form short round sections by clamping a section of pipe in a vise and bending the metal around it. The rounds must be made in very small increments across the whole section of sheet metal. Take your time or you will find yourself unbending your curves because you made them to tight or uneven.

RIGHT ANGLE CYLINDER DEVELOPMENT:

Right angle cylinder development is used to join a burner pipe to the furnace body and to trim the end of a burner to match the inside diameter of a furnace. These surfaces may be developed directly from the layout because all lines are true length or actual size. This is not the case for transition pieces as will be seen later.

To match a cylinder to a specific furnace surface such as the inside diameter:

1. Draw a circle the diameter of the pipe you want to cut.

2. Divide the circle into 12 equal parts (30 degrees each).

3. Extend lines down to D the diameter of the pipe.

4. Draw a circle using the radius of the furnace body (inside or outside depending upon which you are intersecting) so that the outside edge is tangent to the outside edge of the burner pipe.

5. Layout a line that is 12 x L (the length of one section of the burner circle) or you may use π (3.14) times the diameter of the burner.

6. Extend lines up from each of the 12 marks on the layout line.

7. Extend horizontal lines from the intersection of the vertical lines of the burner tube with the furnace radius.

8. Sketch the form shown from the intersection of the lines above the layout line.

9. Add a small tab at one end for joining the cutout section together.

10. Cut out the section.

To properly align the inside and outside sections, use a set off line to locate the furnace radius as seen in the drawing.

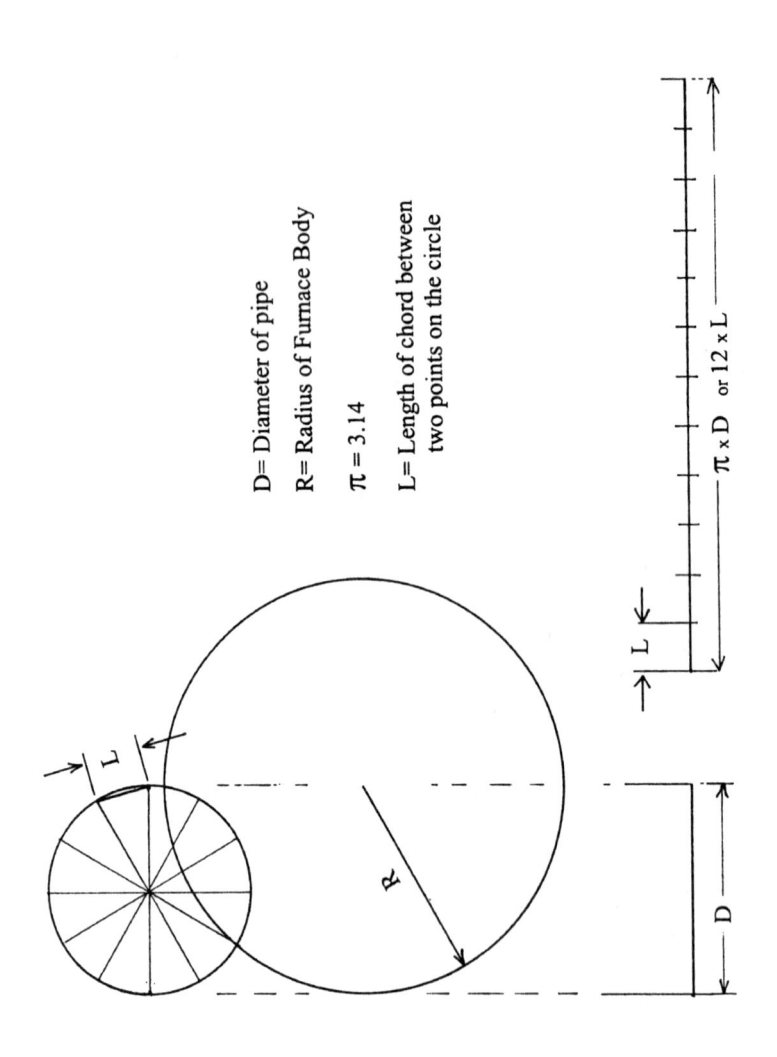

D = Diameter of pipe
R = Radius of Furnace Body
$\pi = 3.14$
L = Length of chord between two points on the circle

$\pi \times D$ or $12 \times L$

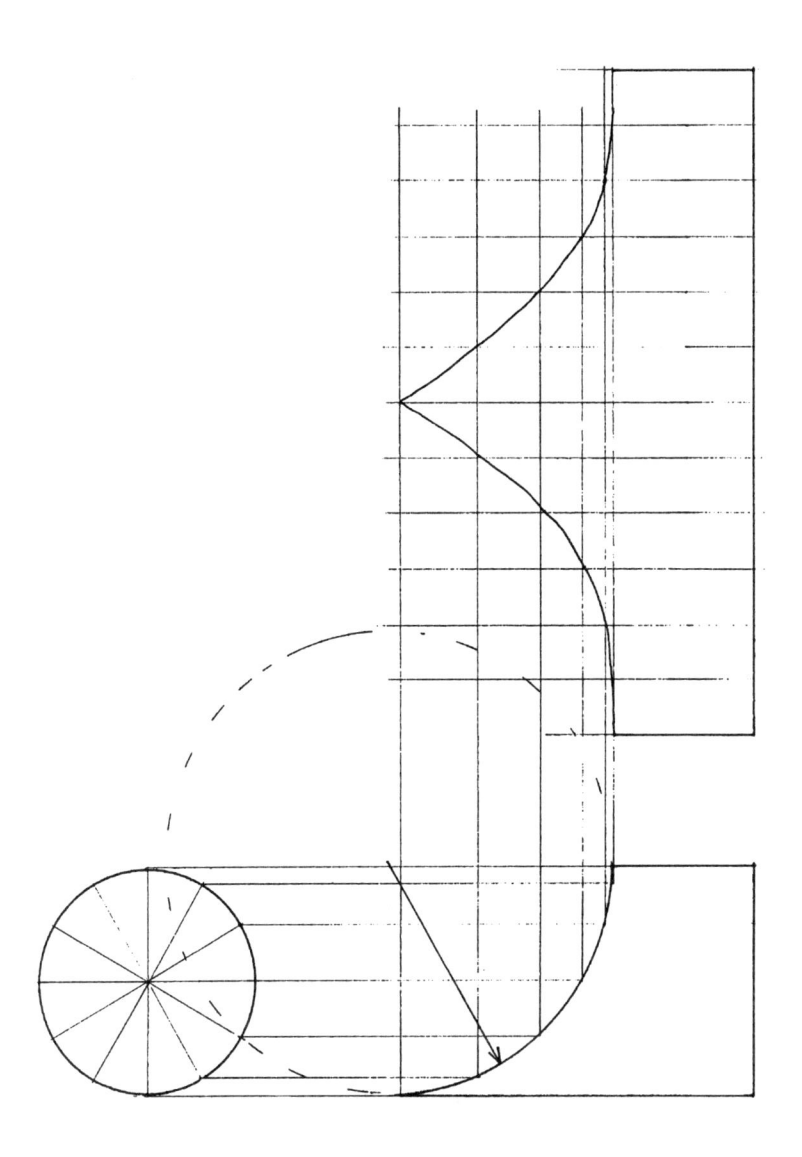

Right cylinders are easy to make because the lines are all true size or actual size. On the transition piece at the end of the chapter, the side and front drawings do not give the true size of all the lines. In this case, the true size lines must be developed by triangulation. It is not difficult to do and may be learned in 30 minutes or less.

MAKING A CONICAL TRANSITIONS PIECE:

When joining two round sections, such as two pipes of different diameter, use a conical transition piece. After the sheet metal has been cut out, it may be formed around a pipe using a rubber hammer or wooden mallet. The steps to make a conical transition piece are given below:

1. Draw a full sized circle the diameter (D1) of the larger pipe.

2. Centered inside the larger circle, draw a full sized circle that matches the diameter of the diameter (D2) of the smaller pipe.

3. Divide the circle into 12 parts.

4. Draw a full size "side view" of the transition piece.

5. Extend the sides to form a complete triangle.

6. Using distance R1, draw an arc.

7. Using R2, draw the larger arc.

8. Using a set of dividers, step off the outer circumference on the larger arc with 12 steps of distance L. This distance will be about 1.1% short, however it can be taken up when you add a ¼ to 3/8-inch tab for joining the sections together. The circumference may also be found by multiplying the diameter by 3.14. You may step off this distance in inches.

9. Draw the sides making sure they intersect at the origin of the radii.

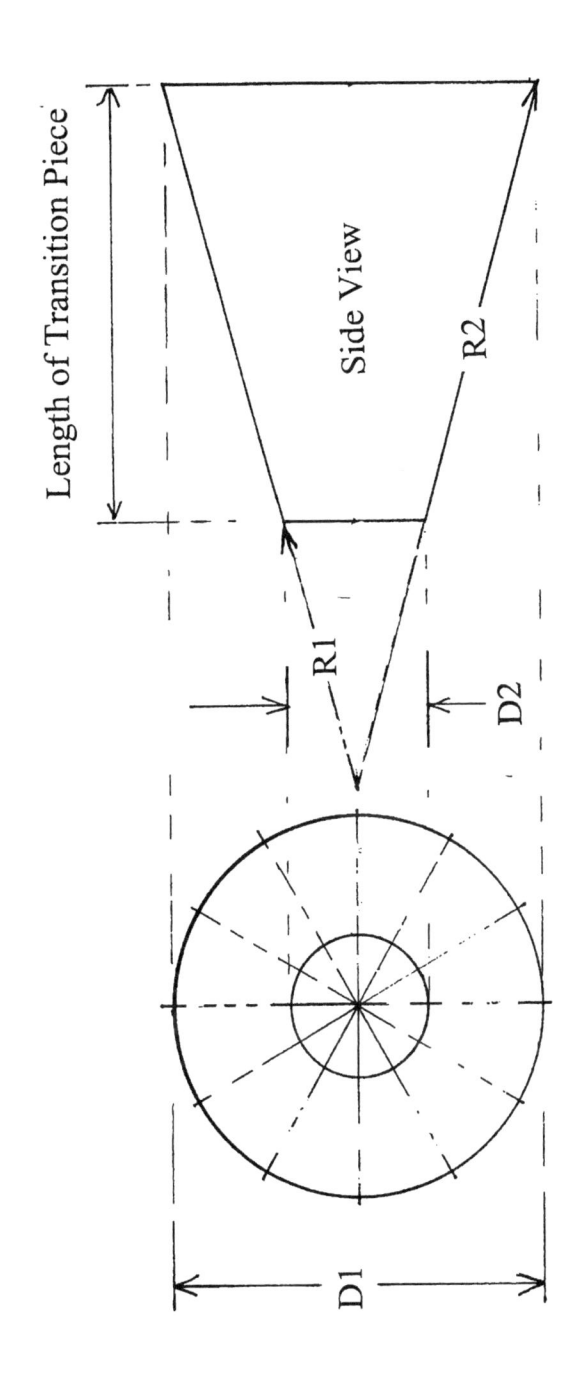

10. Add ½ to ¾-inch tabs on the front and rear. These

should be sectioned for bending around and clamping to the pipes. Add a tab on one side for joining the section together.

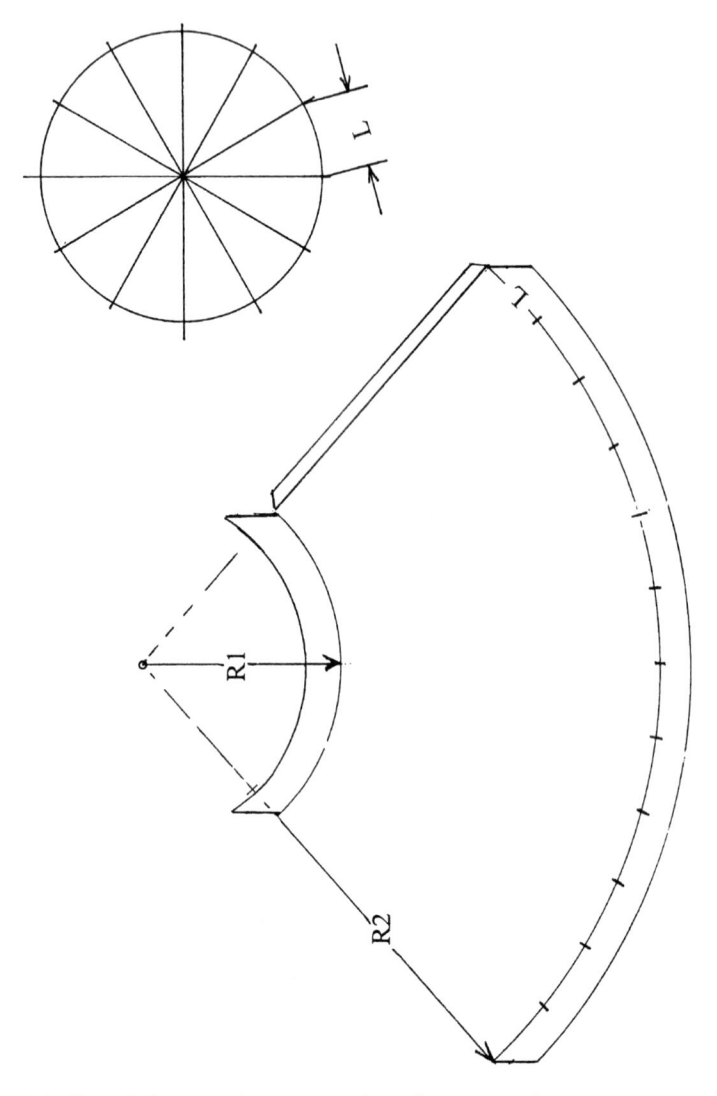

11. Bend the section around a pipe. And fasten together using rivets or sheet-metal screws.

TRANSITION PIECE:

Making a transition piece to go between a rectangular duct and a round duct requires a few more steps than the earlier sheet-metal developments however it builds on the processes already developed. The process is summarized below:

1. Make a full sized drawing of the part.

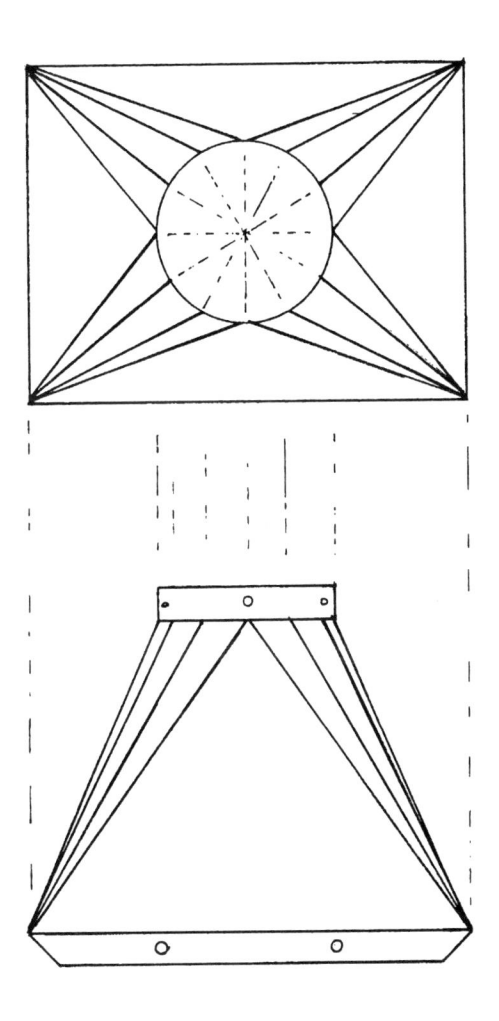

2. Divide the circular duct into 12 parts. Note* *if the round duct is centered* in the rectangular duct, only 4 divisions are required as seen in the next drawing.

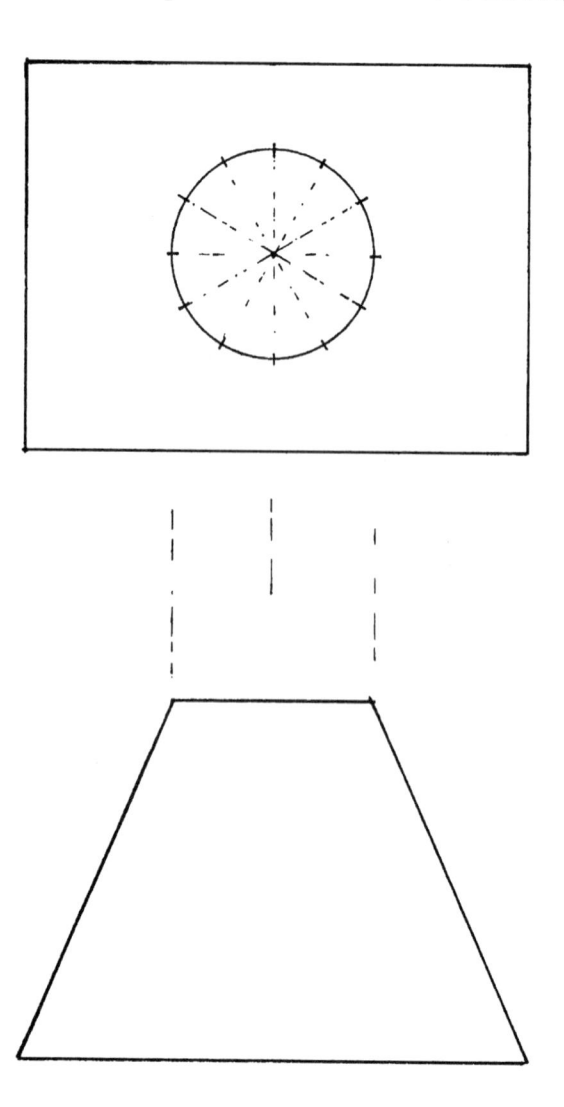

3. Using a compass from point C of the front view, draw an arc from each of the four points 1,2,3,4 to the horizontal line YC.

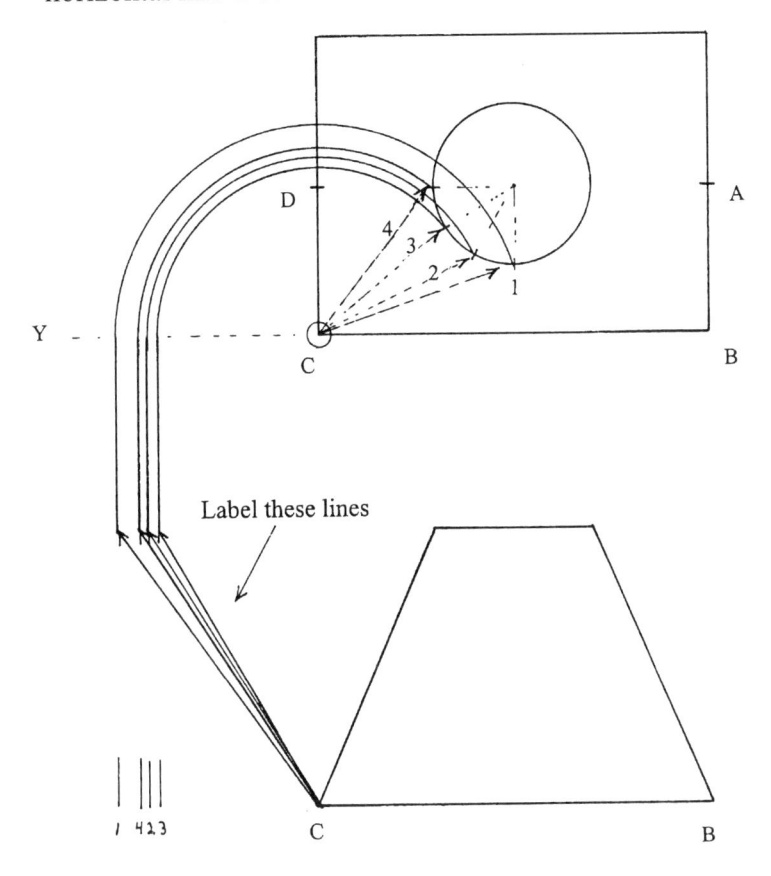

4. Extend the lines down as seen in the drawing.

5. From point C on the side view drawing, extend lines (arrows) to meet the vertical lines drawn in step 4. These lines are the radii that will be used to develop the pattern. They should be labeled 1,2,3 and 4 to prevent confusion when transferring them to the next view. Notice the order of the lines is given at the bottom.

6. Draw two arcs from points C and B using radius 1 from the previous drawing. These will intersect at the first folding lines of the transition piece. Draw the triangle formed by these points.

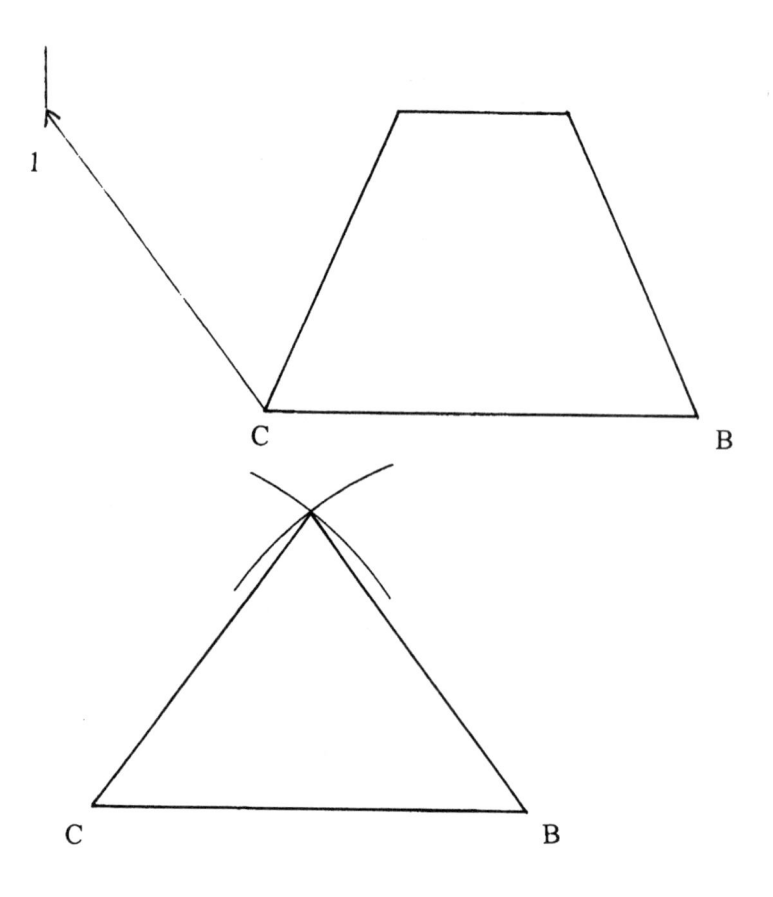

7. Using distance L from the round duct, draw a circle at point X.

8. Using line 2, draw arcs from points B and C so that they intersect the smaller circle. These are the second folding lines of the transition piece.

L

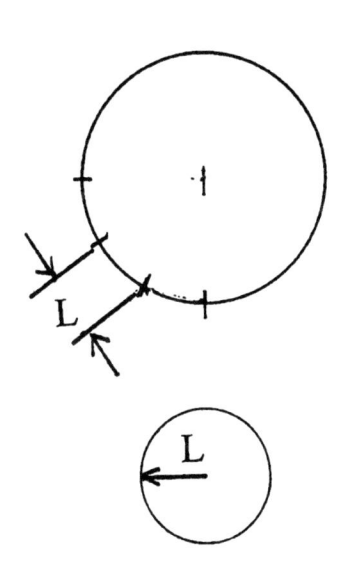

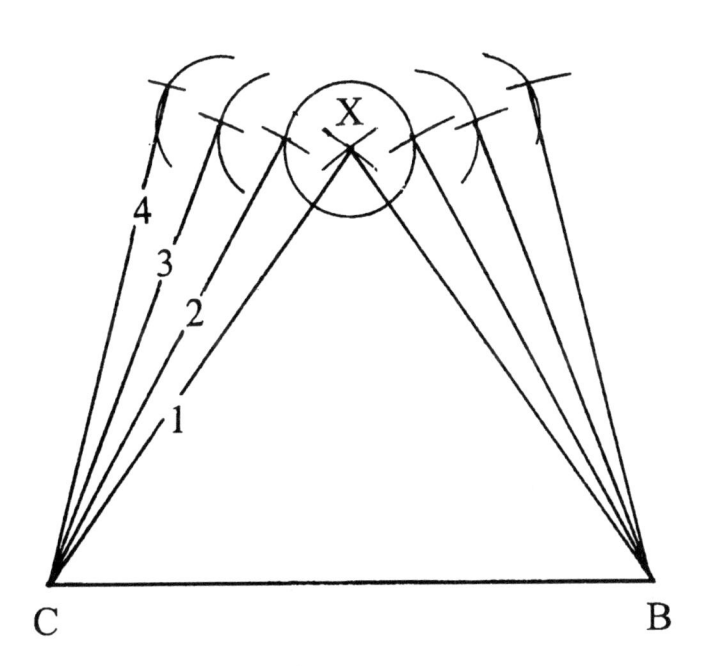

L

4
3
2
1

X

C   B

9. Draw two additional smaller arcs from the previous intersection and repeat the process using line 3.

10. Repeat step 9 using the line 4. Connect the ends with an arc as seen in the next drawing.

11. Transfer lines M and N to the lower drawing to complete *one half* of the transition piece.

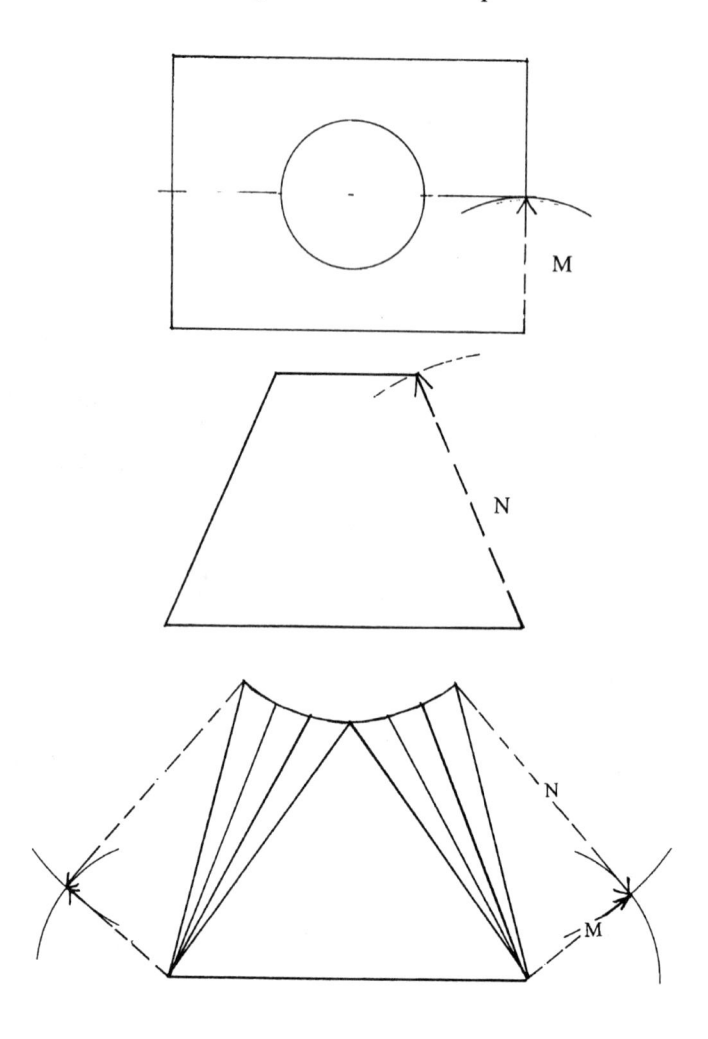

110

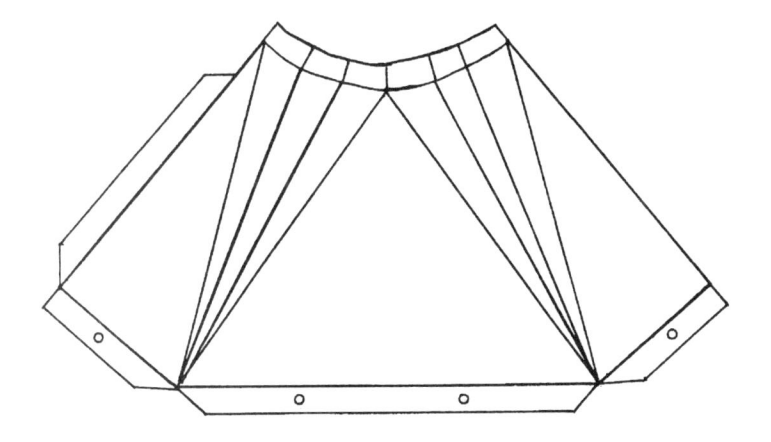

12. Copy the second half of the transition piece and add the mounting tabs.

Completed transition piece on the small furnace blower.

Complete combustion, blower sizing, design, construction and testing are described in "Build a Tilting Furnace" and "Iron Melting Cupola Furnaces."

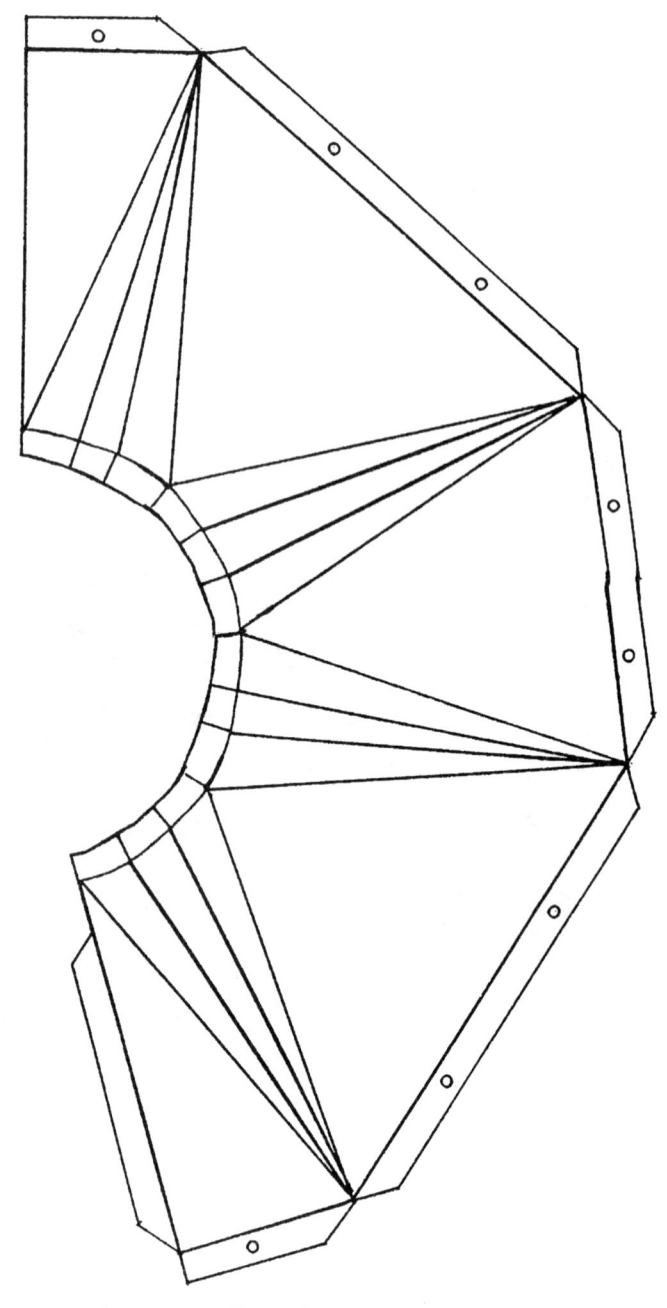

Completed Layout

IV. TEMPERATURE MEASURMENT:

Depending on the range of the temperature and the degree of accuracy required, temperature may be obtained by using a color scale, a thermocouple or an optical pyrometer.

The color scale is the least accurate method of judging temperature for a number of reasons. The ambient light or the lighting of the surroundings produces wide swings in perception of the visible colors. For example, a "dull red" may be seen in a dark room however it might not be visible to the naked eye in bright daylight, or "white hot" in a dark room may appear yellow in bright daylight. Another factor contributing to error in color scale judgments is the variation in color sensitivity of the eye from one individual to another. Equal rates of light input may not produce an equal amount of visual sensation in two individuals. The color scale is included below to give a rough approximation of temperature.

Color	Degrees C	Degrees F
Lowest Visible Red	475	885
Lowest Visible Red to Dark Red	475 to 650	885 to 1200
Dark Red to Cherry Red	650 to 750	1200 to 1380
Cherry Red to Bright Cherry Red	750 to 815	1380 to 1500
Bright Cherry Red to Orange	815 to 900	1500 to 1650
Orange to Yellow	900 to 1090	1650 to 2000
Yellow to Light Yellow	1090 to 1315	2000 to 2400
Light Yellow to White	1315 to 1540	2400 to 2800
White to Dazzling White	1540 or higher	2800 or higher

THERMOCOUPLES:

When two wires of dissimilar metals are joined at one end and this junction is held a different temperature than the remainder of the thermocouple, a voltage is developed across the open ends of the thermocouple wire. The voltage is proportional to the difference in temperature between the junction and the opposite end of the thermocouple.

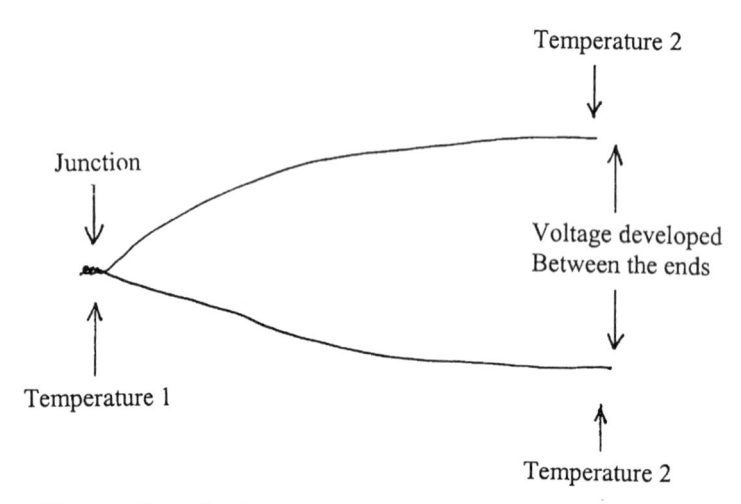

```
                              Temperature 2
                                    ↓
        Junction
            ↓                       ↑
                                  Voltage developed
                                  Between the ends
            ↑                       ↓
        Temperature 1
                                    ↑
                              Temperature 2
```

Four pairs of wire material are most commonly used in thermocouples (meaning there are four common types of thermocouples not four pairs used in one thermocouple). They are:

(1) Type T, Copper (+) against Constantan (-) (60% copper with 40% nickel)

(2) Type J, Iron (+) against Constantan (-)

(3) Type K, Chromel (+) (90% nickel, 10% chromium) against Alumel (-) (94% nickel with 2% aluminum, 1% silicon, and 3% manganese)

(4) Type S, Platinum (-) against Platinum–Rhodium(+) (90% platinum, 10% rhodium, or Type R, (87% platinum against platenium/13% rhodium) Type B, (30% rhodium against platinum with 6% rhodium).

We are primarily concerned with the type K (chromel-alumel) thermocouples in the non-ferrous foundry. Platinum rhodium thermocouples may be used for the higher temperature metals, however optical pyrometers, or infra-red pyrometers may be a better choice for these situations.

Standard tables of thermoelectric emfs (voltages) have been established for these four types of thermocouples. For measuring temperatures in the range –300 to 3100° F, matched pairs of thermocouple wire are commercially available so that they conform to the standard tables. Each strand of wire is calibrated as it is produced. The tolerances are within 0.75% and the precision is good for most technical applications.

Tables are calibrated from the freezing point of water so that the table reads "0 volts" at 32° F (0° C). These tables are generated by having two junctions on the thermocouple. One junction, called the "reference junction," is maintained in an ice bath and the other is heated or cooled while the voltage is recorded throughout the temperature scale.

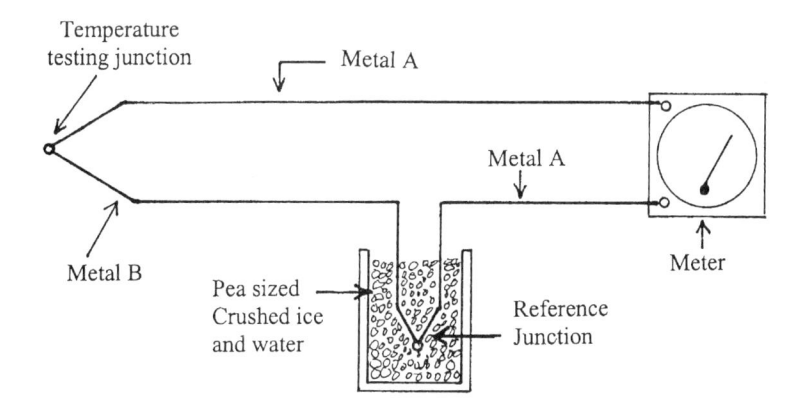

A thermocouple junction temperature may be read from a standard chart by finding the voltage across the open end of the thermocouple.

Example: Find the temperature for a K thermocouple if one end is heated and the other is maintained in an ice-bath if the voltmeter reads 30.49 mV (millivolts).

Reference Junction at 32 F

Degrees F	Millivolts
1310	29.56
1320	29.79
1330	30.02
1340	30.25
1350	**30.49**
1360	30.72
1370	30.95
1380	31.18

The temperature is 1350° F.

Complete Thermocouple Tables are Found in the Appendix

In actual practice, single junction thermocouples are used without the ice-bath. At ambient or room temperature, a single junction thermocouple will read zero millivolts. To correct for the ambient temperature, find the voltage that corresponds to the room temperature and add it to the voltage of the temperature being measured.

Example: If the room temperature is 72° F and the type K thermocouple reads 3.22mV. What is the temperature of the hot junction?

Solution: Reading from the K thermocouple chart at 72° F, you get .88 mV. The sum of 3.22 and .88 is 4.1.

3.22mV + .88mV= 4.10mV

Returning to the table of voltages, 4.1mV corresponds to 212° F, the boiling point of water.

If the ambient temperature is 72° and you insert the thermocouple junction into an ice bath it would read -.88mV.

As you can see, correcting for the ambient temperature is easy to do. Digital thermometers have a circuit to correct for the ambient temperature and are manufactured by Fluke and Tenma among others. They cost approximately $100. The thermocouples supplied with these units should be upgraded for foundry use.

MAKING A THEROMOCOUPLE THERMOMETER:

By using a K type thermocouple, a multimeter and a standard table of thermoelectric voltages, temperatures in the foundry may be measured very inexpensively. A workable system may be assembled for less than $50. By purchasing an 8-inch K type thermocouple probe for approximately $30 from Mifco and using a multimeter from Harbor Freight (cen-tech # P35761) $2.99 on sale (usually $9.99) you can assemble a workable thermometer. You may choose to make your thermocouple probe by twisting and welding the wires together to form a junction, stringing the ceramic insulators and making or buying a protective sleeve; however probes are so inexpensive it is hardly worth it.

Attach the thermocouple tip to a 3-foot section of pipe. You may make a gradual bend in the pipe or you may use a 45° elbow as seen in the drawing. The Mifco probe comes with double bore insulators, yellow and red leads. The probe fits a ¼-inch pipe elbow however it is

118

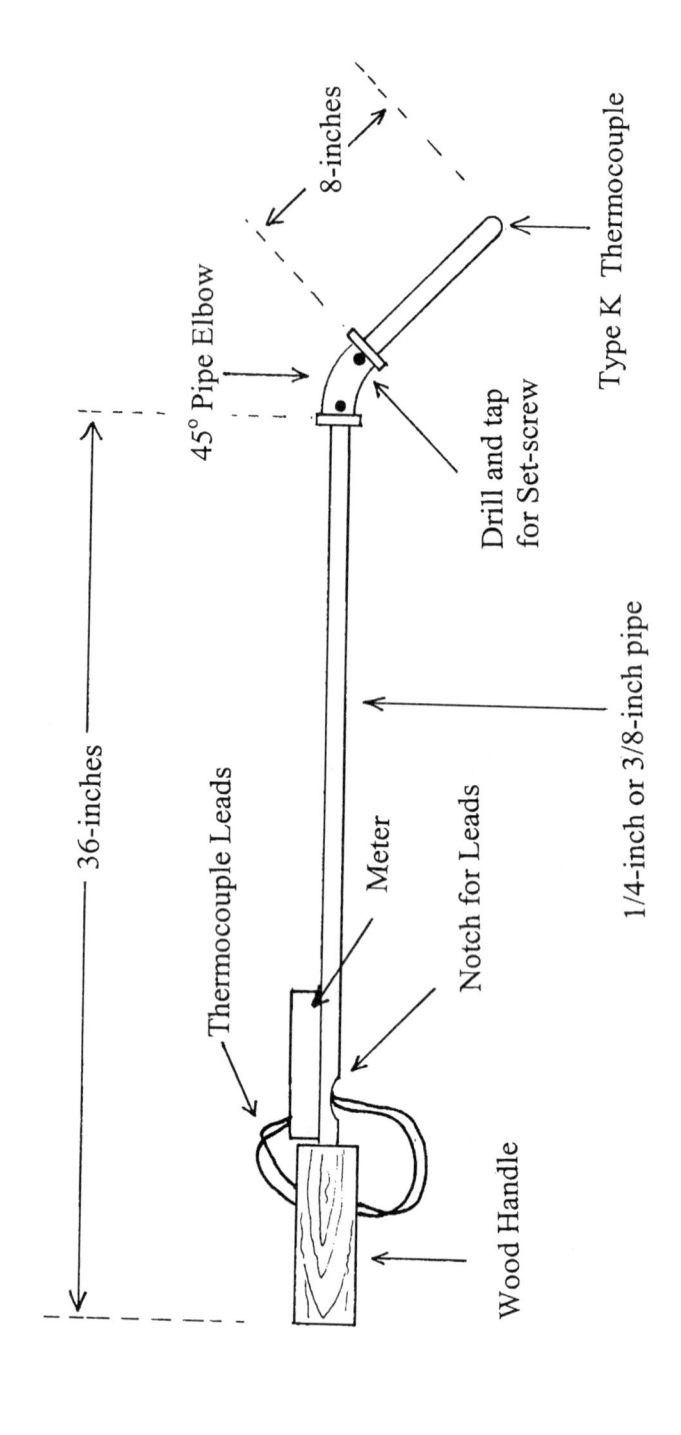

difficult to pull the leads and insulators through the small diameter. Using a 3/8 elbow and pipe make the job much easier. Fish spine insulators may also be used for sharp bends. Make the handle from a section wooden "closet rod" that is available at home building supply stores.

Double Bore Ceramic Insulator

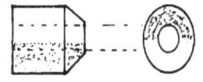

Fish Spine Ceramic Insulator

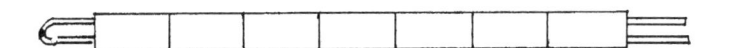

Thermocouple with Double Bore Insulators

| Double Bore Oval Ceramic Insulators ||||
Wire gauge	O.D.	Bore	Length
8	.313 x .575	0.193	1
8	.312 x .500	0.185	1
8	.284 x .500	0.157	3
14-16	.250 x .375	0.090	1
14-16	.187 x .312	0.080	3
20	.118 x .171	0.042	3
All dimension in inches			

Fish Spine Ceramic Insulators			
Wire gauge	O.D.	Bore	Length
8	0.400	0.156	0.400
8	0.260	0.156	0.260
14-16	0.170	0.068	0.170
20	0.110	0.056	0.110

Type K Thermocouple Specifications:

Temperature range for type K thermocouple: -310 to 2500° F

Limits of error for type K Thermocouple: 32 to 530° F + or − 4°F

530 to 2300° F + or -0.75%

Insulation color for type K thermocouple:

Yellow + positive, Chromel

Red − negative, Alumel

Upper Temperature limits for continuous use of protected type K thermocouples by wire size:

No.8	No.14	No.20	No.24	No. 28
0.125 in	.064 in	.032 in	.020 in	.013 in
2300 F	2000 F	1800 F	1600 F	1600 F

The temperatures above usually give a reasonable thermocouple life. For intermittent use, temperatures may be higher.

Protection tubes may be made of ceramic, or silicon carbide. Protection tubes that are used for aluminum melting may be made of cast iron covered with ladle wash. Brass and bronze melting do not require protection tubes.

Temperature Range of Thermocouples in Degrees F	
Type J	up to 1400
Type K	up to 2500
Type R or S	up to 2700
Type B	1600 to 3100

OPTICAL PYROMETERS:

Optical pyrometers allow you to test the temperature of the metal or a furnace without direct contact. Although the unaided eye is not a very accurate method of determining temperature, it can very accurately compare the brightness of one light source to another. In an optical pyrometer you compare the brightness of the hot body with the filament of a lamp. The brightness of the lamp may be changed through a variable resistor or the brightness of the hot body may be changed by viewing it through an "optical wedge." A wedge shaped strip of material similar to the shades in pair of sunglasses is inserted between the hot body and the lamp. The wedge is moved until the brightness of the hot body matches the brightness of the lamp. When the filament of the lamp "disappears" into the hot body, the

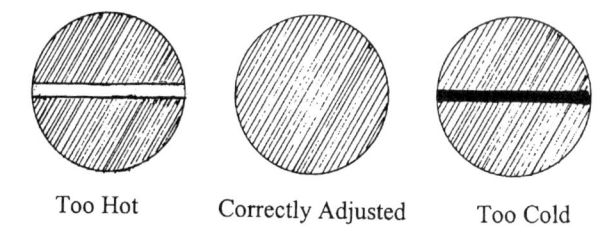

Too Hot   Correctly Adjusted   Too Cold

Disappearing Filament Pyrometer

correct temperature is reached and may be read from a calibrated scale.

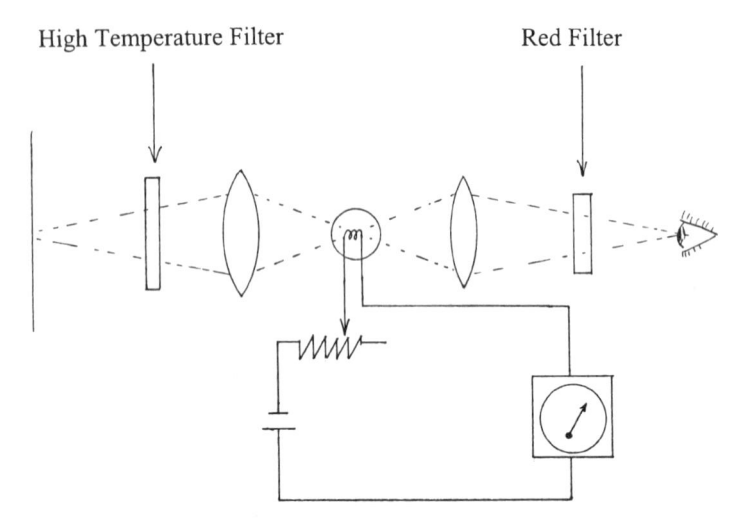

Variable Resistor Pyrometer

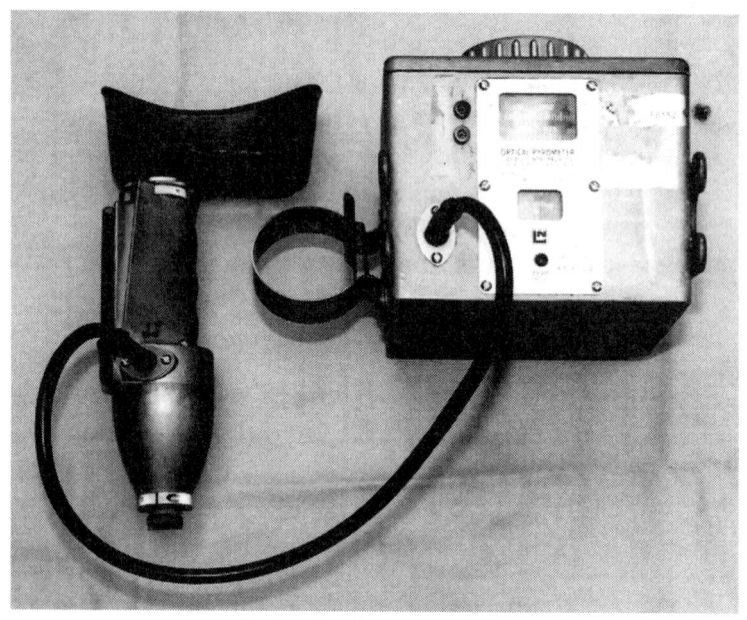

Leeds & Northrup Pyrometer

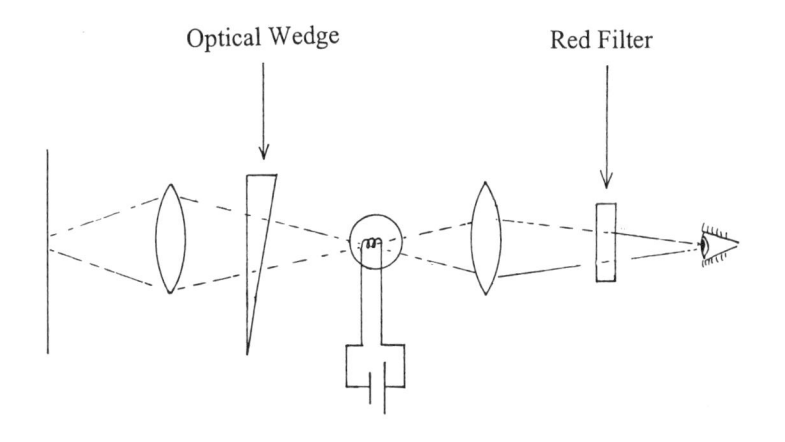

Wedge Type Pyrometer

Pyro Wedge type Pyrometer

Temperatures might be measured using the whole visible spectrum, however it is more accurate to use only one color of light. Red light is used for a number of reasons. 1. At low temperature red is the first color visible therefore readings may be taken at lower temperatures. 2. Red radiation varies twelve times as fast as temperature. Therefore a small change in temperature creates large change in the brightness of red light. 3. Optical pyrometers may be accurately used by those with varying degrees of colorblindness. A red filter is used between the eyepiece and the lamp to block the other colors of light. When the metal temperature is higher than the maximum temperature of the lamp (over 3400° F), a shaded screen is inserted between the hot body and the lamp to reduce the amount of light entering the pyrometer. The temperature is then read from a recalibrated scale.

Optical pyrometers usually have at least two scales, one for black body or ideal emissivity and one for partial emissivity. In general, reflecting or white objects do not easily absorb radiant energy while rough, black objects absorb radiant energy easily. By placing your hand on the roof of a white car on a sunny day and placing your hand on the roof of a black car you will see that the black car will be much hotter. Reflecting and brightly colored objects cool by radiation much more slowly than rough black objects. A polished silver bar cools by radiation much more slowly than a flat black bar because it radiates energy 1/20 as fast as the black bar. Generally, *good absorbers are good radiators*. An ideal absorber and radiator is called a *black body*. The emissivity of a black body is 1. The interior surfaces of a stabilized furnace and rough, oxidized or scale covered metal are nearly black bodies. Molten iron is not a black body and has an emissivity of 0.4. To correct for the difference in emissivity of an object the optical pyrometer used in the foundry has different temperature scales, One for black bodies and one calibrated for emissivity of 0.4. A short table, Emissivity of Materials, is shown on the next page.

The difference in emissivity between liquid aluminum and aluminum oxide may be seen in your crucible. A cover of dross over the metal may have a red tint, however when scraped away, the liquid aluminum underneath is shiny silver with little or no visible red.

Emissivity of Materials: The values given are for red radiation of 0.65 microns wavelength. The left side of the table is for emission from plane polished or unoxidized liquid metal surfaces. In the table on the right, the higher values are from rough surfaces.

Material	Solid state	Liquid state	Material	Range of observed values
Aluminum	0.06	0.12	Aluminum Oxide	.22 to .44
Carbon	0.8 - 0.93		Copper oxide	.60 to .80
Chromium	0.34	0.39	Iron Oxide	.63 to .98
Copper	0.10	0.15	Tin Oxide	.32 to .60
Iron	0.35	0.37	Porcelain	.25 to .50
Cast iron	0.37	0.40		
Steel	0.35	0.37		
Manganese	0.59	0.59		
Nickel	0.36	0.37		
Tungsten	0.43			
Platinum	0.33	0.37		
Silver	0.07	0.07		

V. FOUNDRY SANDS and BINDERS:

MOLDING SAND:

> The Fundamental Foundry Rule: *Use the weakest, driest and finest sand that makes the best castings.* This can only be determined by careful inspection of the castings.

There are two types of molding sand, natural and blended. Blended sand may also be called compounded formulated or synthetic sand. Natural sand consists of silica sand with enough clay bond to be used as it comes straight from the ground. Blended sand is made by adding a bond to a "washed and dried" or unbonded base sand. This unbonded sand is sometimes called "sharp sand," however the name has nothing to do with the actual grain shape. The unbonded base sand may be **silica, olivine, zircon, Hevi Sand** (a chromite sand) or **silicon carbide**. Each base sand has properties that make its use desirable in certain situations. Silica (SiO_2) forms 60% of the earth's crust and is therefore cheap and plentiful. Silica is the most widely used sand. Silica has many forms or phases; however quartz sand is the most common.

Olivine is a solid solution of magnesium oxide 49%, silica 42% the rest being iron and aluminum oxides. Olivine is often used as a base sand. In the United States it is mined in North Carolina and Washington State. Other deposits are mined in Norway and Russia.

Both zircon sand and Hevi sand are expensive and are used mostly as facing sand or mold coatings. Because of the high thermal conductivity of zircon and Hevi sand, they are often used as chill sand to help establish directional solidification.

PROPERTIES OF MOLDING SAND:

For molding sand to be useful it must have certain properties or qualities.

Green Strength is the ability of the sand to hold the shape of the pattern yet allow it to be withdrawn without breaking the mold.

Dry Strength allows the binder to hold the shape of the mold together as hot metal flows over and dries out the surface of the mold. The binder must resist erosion by the stream of hot metal, but not be so strong that shake out or removal of the casting becomes a problem.

Wet Strength: When hot metal is first poured into a mold, the mold surface dries out and the moisture is forced into the surrounding sand. This moisture starts to condense about 3/8 to ½-inch behind the mold face creating an over-wet area of sand. The sand must be able to handle this extra moisture without dropping out into the mold cavity.

Flowability is the ability of the sand to fall into place around a pattern and the ability to flow around a pattern surface to fill the pattern recesses. If sand has good flowability, then it easily forms itself around the pattern when rammed. Sand that is too stiff causes the sand to be forced around the pattern resulting in much extra labor. Factors affecting flowability are the type and amount of bond, the amount of temper water, the sand grain distribution and the amount of mulling. Pure dry silica sand has the highest flowability. As clay and water are added, flowability decreases. Some sand technologist use very small additions of ordinary dish soap to increase flowability.

Permeability: Steam and gas are generated after a mold is poured. These gases must vent from the mold if they are not to be trapped in the casting as bubbles. The sand's ability to vent this gas is called permeability. This gas

Four Types of Sand Grains

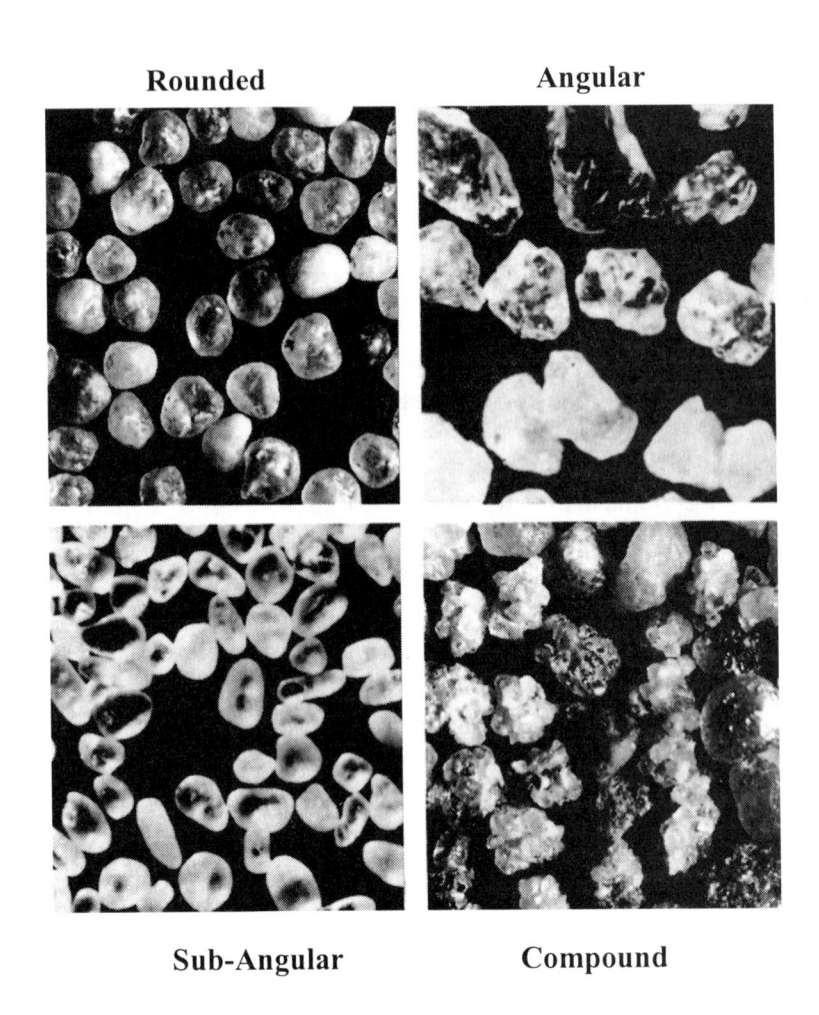

Rounded

Angular

Sub-Angular

Compound

passes between the grains of sand so the size of the individual sand grains, the number of different sized sand grains and the distribution of sizes all affect the sand's permeability. The shape of the sand grains also affects the porous nature of the sand. Clay and smaller sand grains fill the gaps between the larger grains reducing the space available for the gas to escape. Permeability requires open sand, however good surface finish requires that the mold surface be smooth and free of voids between the sand grains. As you can see, good sand requires a compromise between conflicting properties.

SHAPE OF SAND GRAINS: There are four types of sand grains seen in the foundry. They are rounded, angular, sub-angular and compounded. Generally, rounded grains have the least amount of contact between grains, require the least amount of bonding material and have the greatest flowability. They have lower strengths and good venting properties. Rounded sand is often used for cores. Angular grains have good permeability. Because of the interlocking nature of the angular grains, they resist compaction. Fine angular sand is often used in aluminum foundries for good surface finish because of its good permeability. Angular grains require more work to coat with clay because the clay does not smear from grain to grain as easily. They are less subject to metal penetration and washing of sand grains from the mold surface. The properties of the other sand grains are between these two, however the compounded sand grains may break down under the heat of the metal causing a build up of fines and unpredictable molding properties. Compounded grains are less desirable and should be avoided.

SIZE OF SAND GRAINS: Sand is sized by passing it through screens with progressively smaller holes. The hole size is set in inches or millimeters with the finest mesh being #270 at .0021-inches diameter or .053 mm. Screens progress by a ratio of $\sqrt{2}$ or 1.414 times the previous diameter. The next size screen is #200 so the hole diameter is 1.414 x .0021-inch = .00296-inch. Screen sizes are summarized below.

US Screen Sizes	Tyler Series meshes per inch	Openings inches	Openings mm	Diameter of wire inches
4	4	0.1870	4.699	0.0650
6	6	0.1320	3.327	0.0360
8	8	0.0937	2.362	0.0350
12	10	0.0661	1.651	0.0320
16	14	0.0469	1.167	0.0250
12	20	0.0331	0.833	0.0172
30	28	0.0232	0.589	0.0125
40	35	0.0165	0.414	0.0122
50	48	0.0117	0.295	0.0092
70	65	0.0083	0.208	0.0072
100	100	0.0059	0.147	0.0042
140	150	0.0041	0.104	0.0026
200	200	0.0029	0.074	0.0021
270	270	0.0021	0.053	0.0016

THE EFFECT OF SMALLER GRAIN SIZE:

The surface area of a given weight of sand increases rapidly with a decrease in size of the grains. To illustrate the rapid change in surface area the following example is presented:

Number 20 screen sand is .0331-inches in diameter. If a 1 inch cube is divided into smaller cubes that are .0331-inch long on each side you will get 27,575 smaller cubes.

1-inch^3 / (.0331-inch)^3 = 27,575

The surface area each small cube is .00657 square inches.

(.0331-inch)^2 x 6 sides of a cube = .00657 square inches

The total surface area of the smaller cubes is

27,575 x .00657 square inches = 181 square inches

The surface area of a single 1-inch cube is 6-inches. If these calculations are repeated for a number 270 screen, the surface area of the small cubes increases to 2857 square inches. As you can see, the surface area increases rapidly with smaller sand size. The relationship of progressively smaller cube size to surface area is shown in the table below.

Length of side Inches	Number of cubes	Surface Area in Square Inches	Surface area in Square Feet
1	1	6	0.0417
0.1	1,000	60	0.4167
0.01	1,000,000	600	4.1667
0.001	1,000,000,000	6,000	41.6667

All sand properties are probably dependent or associated with the surface area of the base sand grains. The amount of binder required is directly proportional to the grain size. Because the surface of each grain must be

Grain Size	Surface Area sq in./ lb
40	4,300
50	6,100
70	8,600
100	12,000
140	17,000
200	24,000

covered with binder, the smaller grain size with its greater surface area requires more binder.

The effect of fines on permeability is significant. Permeability drops quickly as the grain size is reduced, therefore fine sands have low permeability. Coarse grains have larger void spaces between the grains that increases permeability. For grain sizes from 140 to 270 the permeability is very low; however permeability increases rapidly as the grain size goes from 140 to 20.

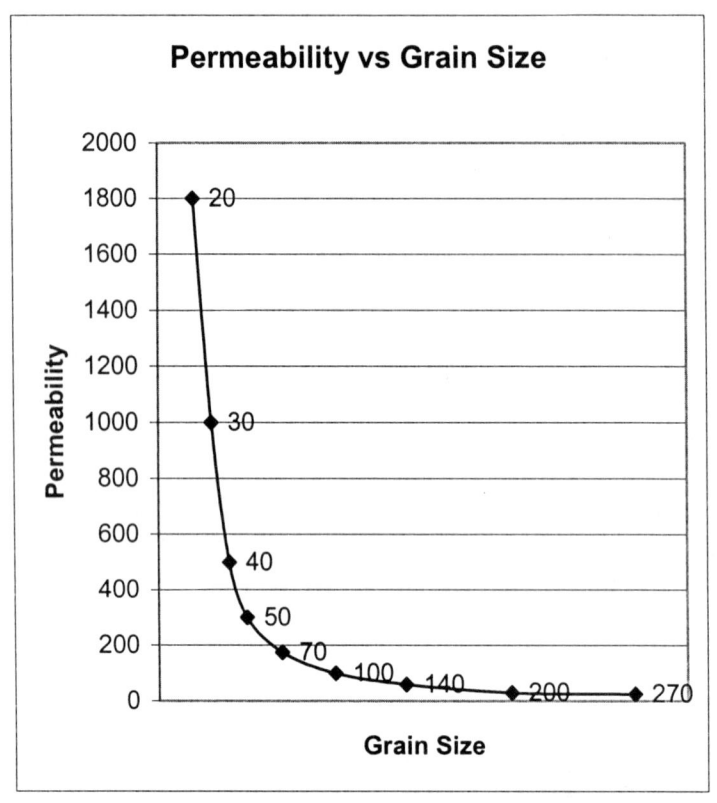

Permeability vs Grain Size

Grain Size	Permeability
20	1800
30	1000
40	500
50	300
70	170
100	100
140	—
200	—
270	—

Grain size distribution is also a major factor in permeability. Sand with a mixture of several grain sizes and many fines has a low permeability when compared to sand of one-grain size. This is because the smaller grains fill the voids between the larger grains as seen below.

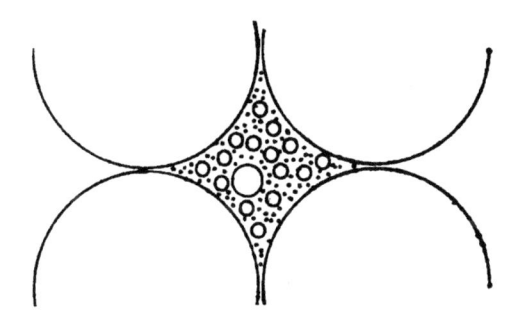

When there are high percentage of fines in the sand, the smaller grains fill in the spaces between the larger grains, reducing the permeability.

GRAIN FINENESS NUMBER:

The American Foundry Society Grain fineness number is the *average* grain size in a sample. It is calculated by passing a 50-gram sample of dry sand through a series of screens and weighing the amount caught on each screen. Multiply the weight in grams on each screen by two in order to get the percent present in the sample. The percent is multiplied by the figure as seen in the table below. For instance, 16 grams of #70 sand are found on the screen:

16 x 2 x 50 = 1600 as is seen in the example table. Add up all the numbers in the product column. This equals 6500. Then divide by the total percent retained on all the screens.

AFS# = (total product) / total percent retained

AFS# = 6500 / 100 = 65

Because the AFS grain fineness number is an *average*; sands with the same AFS number may have very different screen analysis. *Two sands with the same AFS number may produce different surface finishes.* **The actual distribution of the sand grains is more useful than the AFS grain number.**

Example:

US series sieve	Grams	Per cent	Multiplier	Product
6	0		3	
12	0		5	
20	0		10	
30	0		20	
40	0.5	1	30	30
50	10	20	40	800
70	16	32	50	1600
100	14.25	28.5	70	1995
140	6.75	13.5	100	1350
200	2.5	5	145	725
270	0		200	
pan	0		300	
Total	50	100		6500

SAND GRAIN DISTRIBUTION:

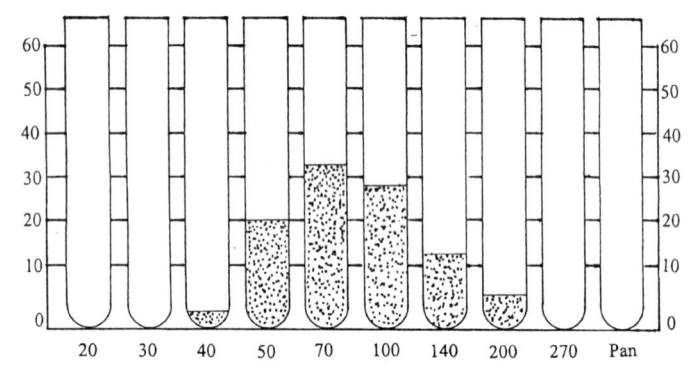

An engineer might want to discuss sand in terms of the AFS number, but many "practical foundrymen" would rather use a grain distribution such as a three-screen sand or four-screen sand to identify the sand. A screen fraction is a screen that has more than 10 % of the total sand caught on it. The AFS# 65 in the example above is a 4-screen sand. This is a good sand for hotter metals and larger castings.

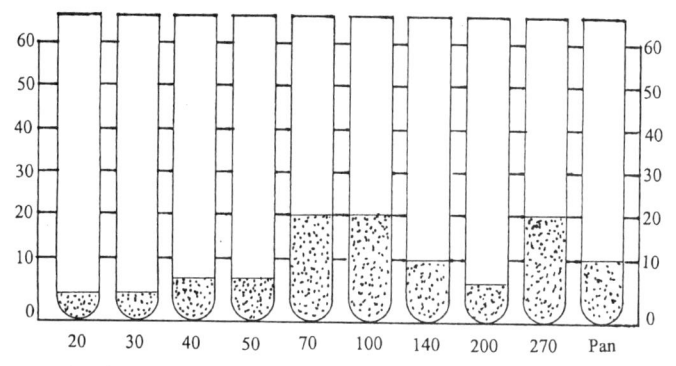

The sand above has too much fine material. Due to the surface area of the fine sand, it would require much bonding material and soak up large amounts of water. Pin holes are common in castings made in sands that have large amounts of temper water.

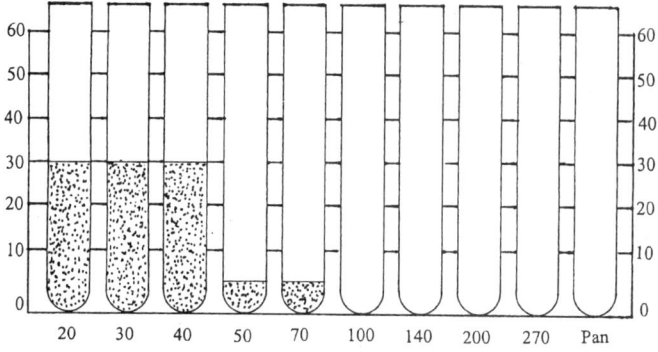

This sand is too coarse. It is brittle and produces poor surface finish due to metal penetration between the sand grains.

NATURALLY BONDED SAND:

Useable natural molding sand can be found in many parts of the country. Natural molding sand may be found along the banks of a river or in a creek bed and in dunes, among other places. Bank sand often has an overburden that has to be removed. While in Georgia, I found good sand on the side of a hill that was being cut away for a new road. Albany sand, one of the most popular natural molding sands, comes from Albany, New York. It is used when an extremely fine surface finish is required.

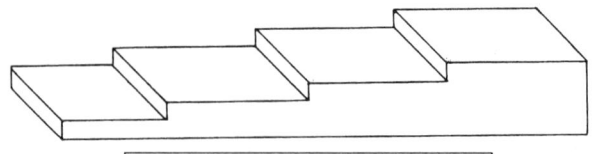

Test Wedge ¼ to 1-inch thick

Natural sands often contain lime, iron oxide, and alkali oxides like that lower its fusion temperature. At higher temperatures the sand may fuse into glass that sticks to the casting. Because of its lower fusion temperature, natural molding sand is primarily used for casting lower temperature metals such as aluminum, smaller brass and bronze. Higher temperatures and heavier castings require a higher fusion point sand. You can test naturally bonded sand by pouring a test casting as illustrated. After pouring, clean the casting with a wire brush and look for sand that has fused to the casting. If the sand fused to the ½-inch section, then your castings should be less than ½-inch thick. If the test wedge is clean up to the 1-inch thick section where sand starts to fuse to the casting, then your castings should be less than 1-inch thick. If no sand has fused to the casting, then you may want to pour another thicker test wedge to determine the maximum casting thickness the sand will support. Sand does not fuse at the temperatures that aluminum is poured. The wedge test is used for brasses, bronze, iron and steel.

COMPOUNDED (SYNTHETIC) SANDS:

Compounded sands are used over natural sands because they may be formulated by using a base sand of proper size and adding a binder in proper proportions to make the best sand for a particular type of casting.

TYPES of BASE SANDS: There are several types of base sands available to the foundryman. They include **silica,**

olivine, zircon, Hevi sand and **silicon carbide.** Each sand is selected for its specific properties.

Silica sand is the most common of all foundry sands. It is a good refractory, cheap and plentiful however it has one bad trait. It has large volume changes with changes in temperature. Quartz crystals change shape at definite temperatures. Up to 1063° F (low) quartz has one crystal shape. At 1064° F it makes an abrupt change to high quartz with a 1.5% increase in volume. At 1598° F it changes to tridymite. At 2678° F it changes again to crystolbalite.

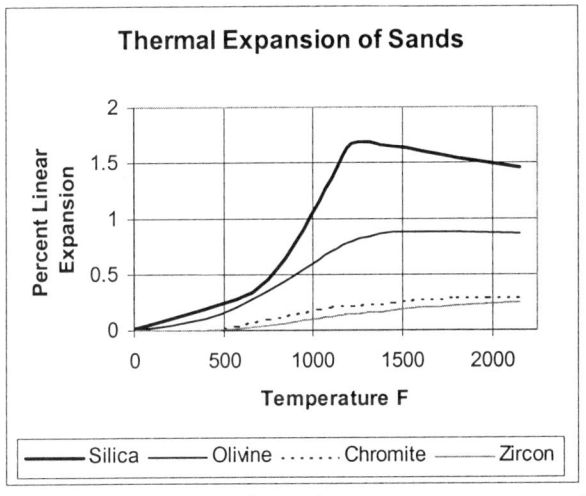

Thermal Expansion of Sands

(Graph: Percent Linear Expansion vs Temperature F, showing curves for Silica, Olivine, Chromite, Zircon)

These abrupt changes in volume cause stresses in the mold. During the first stages of expansion, the cope sand may rain from the mold surface into the flowing metal causing sand inclusions in the casting. Rattails, buckles and scabs may also form. The sand's expansion must be balanced by contraction in the clay, wood flour or other organic material added to the sand. Venting the cope allows areas of movement as well as venting the escaping gasses. "Burned in" sand or sand that has been used a few times has some permanent expansion in the crystal structure. This used sand is less likely to cause scabs rattails and buckles and is preferred to new sand. However, aluminum does not get hot enough to permanently change the quartz crystals into a more stable form. The sand expands and contracts back to its original shape with each heating and cooling cycle.

Compacted silica sand weighs about 98 pounds per cubic foot. Silica has a specific gravity of 2.65.

SAND MIXES:

Hand mixed sand for small iron and bronze castings*:

Note: Because it is a hand mixed sand, the clay contents are higher. The fireclay addition gives a high hot strength. In order to reduce the hot strength, 1 pound of fine sawdust may be added for cored brass. This sand may use a plumbago wash sprayed over the mold surface. *Recommended by Stewart Marshall

Fine silica	100 pounds
Southern Bentonite	10 pounds
Fireclay	15 pounds
Seacoal	3 pounds
Temper water	7 to 10 pounds

A local foundry has been casting Aluminum and bronze for about 50 years using the mix shown below:

150 pounds #120 silica sand
 25 pounds #180 silica sand
 8.5 pounds western bentonite
 1.5 pounds southern bentonite
 3 pounds water

Other Silica Sand Mixes:

Aluminum Mix (by weight):	
Silica sand AFS # 90 -180	95%
Southern Bentonite	5%
Temper water	3%

Gray Iron Mix – Light Castings (by weight)

Bank Sand (silica) AFS # 90 to 130	89%
Southern Bentonite	6%
Seacoal	5%
Temper water	3-3.5%

Gray Iron Mix – Medium Castings (by weight)

Silica sand AFS # 60	91%
Western Bentonite	4%
Sea coal	3%
Southern Bentonite	1%
Carbonized Cellulose	1%
Temper Water	3-3.5%

Light Bronze and Brass Mix: (by weight)

Silica sand AFS# 60-90	50%
Silica sand AFS# 90-120	43%
Western Bentonite	2%
Southern Bentonite	3%
Sea Coal	2%
Temper Water	3%

Light Bronze and Brass (by weight)

Silica sand AFS# 90-120	73%
Naturally bonded sand AFS# 250	25%
Southern Bentonite	2%
Temper water	5-6%

Bronze-Small to Medium-Flat Castings (by weight)

Silica Sand AFS #115-150	57.5%
Albany Naturally Bonded Sand	40.0%
Southern Bentonite	2.0%
Wood Flour (finer than AFS#50)	0.5
Temper Water	5.0%

Steel Sand – Small Castings: (by weight)

Silica sand AFS# 60	64.5%
Silica sand AFS #90	30.0%
Western Bentonite	5.0%
Corn Flour	.5%
Temper water	3.2%

Olivine sand is 25% heavier than silica sand and weighs about 122 pounds per cubic foot. It has a specific gravity of 3.31. Olivine has a sharply angular grain because it is mined as rock and must be crushed before use. The angular grain allows the use of finer sands while keeping good permeability. Molds can be rammed harder. Olivine has a lower thermal expansion rate than silica sand. Olivine has a thermal expansion of .0083 inches per inch. It has no sudden expansion as seen in the phase changes of silica sand, therefore it has lower scabbing and buckling tendencies. There is little need for the addition of wood flour. Olivine has very little free silica therefore does not cause silicosis or a similar condition. Other properties of olivine are high heat absorption and conductivity below $1200°$ F temperatures. It acts as an insulator at high temperatures.

Non-Ferrous Molding Sands

Mixture	A	B	C	D
#130 olivine	94	95	94	74
#200 Olivine				20
Western Bentonite	3	4	1	1
Southern Bentonite	3	1	5	5
Water	3.1	3.2	3	3.5

Mixture A in the table above is a general purpose sand for aluminum and brass work. It is a good starting mixture that may be modified to suit particular needs. Mixture B is used for brass and bronze work up to 500 pounds. Mixture C has a higher green strength. Mixture D is for small thin sections that must be accurate and have a smooth surface finish.

Gray Iron Molding Sands: Most iron sands are based on a #70 grain bonded with western bentonite. Addition of #120

Mixture	A	B
#70 Olivine	0	94
#120 Olivine	95	0
Western Bentonite	5	4
Southern Bentonite	0	2
Water	3.5	3.1

for a smoother finish is common practice Mixture A is used for small gray iron or facing sand on large gray iron castings. Mixture B is used on larger iron castings.

Zircon Sand or zirconium silicate is principally found in Florida beach sands however the concentration is low, being only about 2%. Australian beach sands are approximately 15% zircon. Zircon is expensive because it must be concentrated, washed and graded before it is useful to the foundryman.

The average density of zircon is 172 pounds per cubic foot. It has a specific gravity of 4.7 and a thermal expansion of approximately .003 inches per inch, which is less than one third that of silica. At 2000° F Zircon has 1/6th the expansion of silica. Zircon sand has round grains. Because of its high heat conductivity, it may be used to chill castings. It has a thermal conductivity about four times greater than silica sand. Casting surfaces are good when using zircon sand because it is not easily wet by molten metal. Metal flows across it like mercury on a glass plate. Prepared Zircon mold washes are available and commonly used.

Heavi sand or chromite spinel sand is largely produced in South Africa. It must undergo an extensive separation process before it is useful to the foundryman. Hevi Sand weighs about 160 pounds per cubic foot and has a specific gravity of 4.5. It has a fusion point of approximately 3800° F and a thermal expansion of .005 inches per inch. Heavi Sand has high chilling properties (4 times that of silica) that may be used to promote directional solidification at corners or sharp angles. Hevi sand is used primarily for large castings or intricate cores that are subject to burn on, metal penetration, scabbing or erosion.

Silicon Carbide Sand is produced in an arc furnace by heating sand and coke in the presence of sawdust with salt as a fluxing agent. Its high cost prevents it from being widely used in the foundry. Silicon carbide must be crushed, therefore it has an angular grain. It has a specific gravity of 3.2 and withstands temperatures of $4200°$ F. It is used where high heat transfer is required for chilling castings. Sometimes fine metal shot or grit is added to silicon carbide molding sand to further increase the chilling tendencies.

SAND ADDITIVES:

The best sand mixes are the simplest ones. If you are producing good castings with sand, clay and water, there is little need make additions to your sand. There are many types of additives, however the main functions are to increase the green strength, to provide collapsibility for the expansion of the silica grains, to improve surface finish of castings and to improve the density of the sand to resist metal penetration.

Wood flour, ground corn cobs or other type of cellulose may be added to balance the high temperature expansion of the sand by burning out at $400°$ F to $600°$ F. It reduces hot tears and cracks. Cellulose improves shakeout by making the molds more collapsible and improves the flowability of the sand mixture. It is usually added in amounts of .5% to 2%. Excessive additions may cause rougher casting
surfaces because the metal penetrates the voids left between the sand grains as the cellulose burns out. Brittle sand, crumbly mold edges and gas defects are also common with excessive additions of cellulose.

Seacoal or coal-dust is added to improve the surface finish and cleaning of iron castings. Sands peel freely from castings when sea coal is used. Seacoal should be ground to the same size as the base sand in order not to change the permeability. As the metal flow across the sand and the temperature of the sand increases, some of the seacoal starts to coke releasing light oils and tars. These form a soot or carbon film over the mold and help prevent the sand

Summary of Base Sand Characteristics:

	Silica	Olivine	Zircon	Chromite	Silicon Carbide
Weight in pounds per cubic foot	98	122	172	160	
Specific Gravity	2.65	3.31	4.7	4.5	3.2
Thermal Expansion	0.018	0.0083	0.003	0.005	
Heat Transfer	Average	Low	High	High	High
Fusion Point	2600° F 3200°F	2800° F 3200° F	3700° F 4000° F	3600° F 3800° F	4200° F
Grain Shape	Angular to Round	Angular	Round	Angular	Angular
Wettability	Easily wetted	Not Generally	Resistant	Resistant	
Grain Distribution Number of screens	2 to 5	3 to 4	2 to 3	4 to 5	

grains from fusing to the castings. Seacoal burns in the mold consuming the oxygen and producing a reducing atmosphere. Seacoal comes in grades A,B,C, and D with C and D being the finer grades. "Dustless" seacoal has small additions of oil or wax. Seacoal is added from 2% to 8% by weight. Iron sands use about 5% sea coal by weight. Too much seacoal causes veining, fins, and brittle sand. Temper water is increased 10% of the weight of the seacoal.

Silica flour is sand that is finer than 200 mesh. It is added to increase the hot strength of sand and resist metal penetration. Olivine, chromite, and zircon may also be purchased as "flour." Permeability of the sand decreases and the temper water requirements increase due to the increased surface area of the smaller grains. The mold hardness rapidly increases with small additions of silica flour.

Graphite or **Plumbago** is added to sands from .2 % to 2%. Metal does not wet sand mixtures with 1% to 2% graphite. Graphite is often dusted on molds to improve casting finish and is often used with coal. Foundry blackings or mold coatings usually contain graphite.

Cereal may be wheat four, corn flour, or rye flour. It is added up to about 2% to increase the green and dry strength of molding sand. It is also added to core sand up to about 5%. Because it burns out, it also increases the collapsibility of the sand. Cereals may be added to correct brittle sand. The sand grains do not rub off as easily on air-dried molds using cereal binders. Cereal decreases the flowability of sand. It provides a cushion between sand grains that helps control high temperature expansion. Cereals may tend to make molding sand "ball-up."

Dextrin is also a cereal type binder that gives high dry strength to detailed mold edges. It migrates to the mold surface like molasses in a drying mold.

Molasses is diluted 10 to 1 with water and used in core sand mixtures, core washes, blackings mold sprays and as an additional binder in clay bonded sand.

Fuel Oil or **Kerosene** may be added in very small amounts .01% to .1%. It improves flowability, prevents rapid edged drying of the molds and provides a reducing atmosphere. It may replace or reduce seacoal additions.

CLAYS:

Two types of clays are used in the foundry, kaolinites or *Fireclay* and montmorilllonites or *Bentonites*. Bentonite clays include "western or sodium bentonite" and "southern or calcium bentonite." Each type of clay has specific properties that make it useful to the foundryman. Often a combination of clays is used to get the best properties of each one.

Fireclay is found in most naturally bonded sands. It is sometimes used in compounded sands for high hot strength or to make a sand less sensitive to variations in moisture content. Combinations of fireclay and western bentonite can reach hot strengths over 200 psi. Fireclay has only about $1/3^{rd}$ to $1/5^{th}$ of the bonding strength of the bentonite clays. It also requires more water. A fireclay mixture will be approximately 12 % to 15% by weight clay and 5% to 8% water for the maximum strength. Fireclay has a high refractoriness or softening point of $3100°$ F.

Bentonites: Bentonite bonded sands usually have between 3% to 6% clay and 2 ½% to 4% water.

Western bentonite is a swelling clay. It swells from 10 to 20 times its original volume. It has high hot strength that prevents cutting and erosion of the mold as the metal passes over it. Western bentonites have hot compressive strengths of about 80 psi. Western bentonite mixes have less "flowability" or are more gummy and stiffer than southern bentonite sands. The sand is more rubbery and must not be rammed too soft. Because the sand has higher green deformation (rubbery), patterns are easier to lift from sands bonded with western bentonite. Western bentonites have a tendency to form clay balls. The softening point of Western bentonite is $2100°$ F to $2450°$ F.

Southern Bentonite gives good flowability to sand. It has higher green compressive strength than western bentonite but a lower hot strength of about 40-psi. The lower hot strength reduces hot tears or cracked castings as castings cool. Southern bentonite breaks down easily. Shakeout and cleaning of the castings are easier than with western bentonite. The softening point of southern bentonite is 1800 °F+.

BONDING IN SAND-CLAY MIXTURES:

Clay particles, when placed in water become electrically charged. When clays are hydrated, or adsorb water, the water molecules separate into charged particles or ions that attach themselves to the surfaces of the clay particles. The bond between the clay particles and the hydrated quartz particles is due to the electrostatic forces between the hydrated particles. Clays that have smaller particles have a larger surface area and therefore have greater bonding strength.

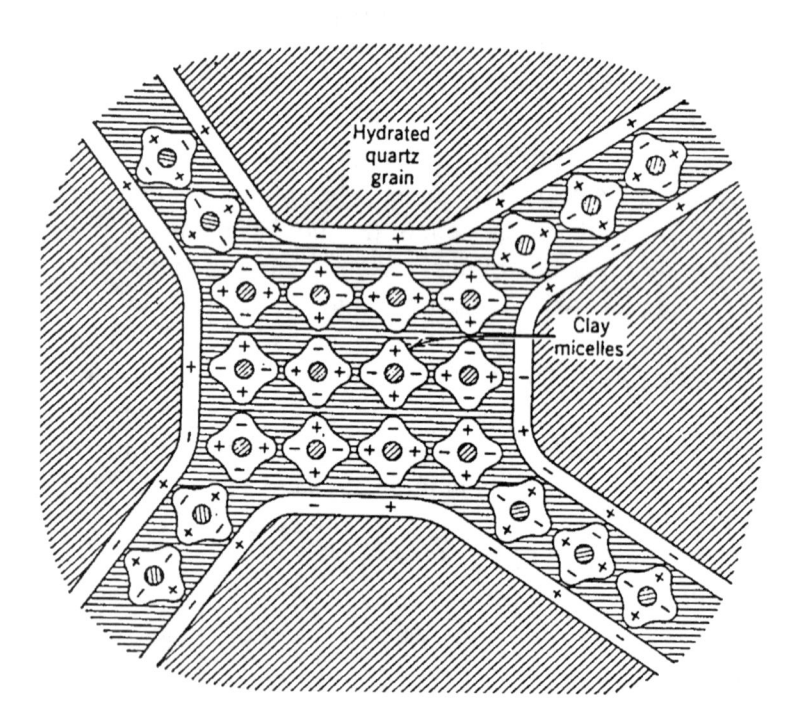

Clays are made up of very small particles that are approximately 1/5000 to 1/50,000 of an inch in diameter and look somewhat like sheets of paper. When dispersed in water, they separate into particles that are so small that many of them are beyond the range of optical microscopes. Although they are extremely small, kaolinite (fireclay) crystals are about 60 times larger than the bentonite clays. Sodium (western) bentonite particles are smaller than calcium (southern) bentonite therefore sodium bentonites have a larger surface area.

Sodium bentonite absorbs water into its crystal lattice. This causes the clay to swell like an accordion. As the clay dries out, the lattice shrinks back to its original size. The shrinking clay helps balance the expansion of silica sand in hot molds. Sodium bentonite forms long knitted chains in the sand mixture. These chains remain intact when the clay dries out giving it high dry strength.

CLAY SATURATION:

Sand mixtures can be classified as clay saturated or unsaturated. Adding more clay to a saturated sand does not increase the green strength. A clay-saturated sand has the same strength as the clay by itself. Bentonite bonded sands (approximately 60 to 100 AFS) saturate at 8% to 12% clay content. Similar fireclay bonded sands saturate at 20% to 25% clay content. Saturated fireclay sands have low permeability and high moisture content and therefore are not commonly used. Clay saturated sands, when hard rammed develop green compressive strengths of 14 to 20 psi.

For any clay there is an optimum combination of clay and water. Too much water causes the clay to be gummy and causes excessive hot strength. Too little water causes the mold to be brittle, weak and causes sand washes in the mold. If your sand is sticking to your rammer, you have too much water.

MULLERS and MULLING:

After you have made a several molds or your operation becomes larger than an occasional small casting, you will be looking for ways to reduce the amount of work in tempering and conditioning the sand. Mullers are used to evenly coat and distribute the clay and water bond to each grain in the molding sand. This is a difficult job because near the ideal temper, the clay is a tough plastic mass resembling modeling clay.

Good sand mulling requires the ingredients to be accurately measured. Water is the most critical of the sand-clay mixture and should be measured exactly. Second, it must be well distributed throughout the batch. Proper mulling time is essential for good mixing. Under-mulled sand causes molding problems so there is no time saved by using short cycles. Six to seven minutes is generally used to develop a good bond.

The water and clay bond must be kneaded into a dough like consistency. The best mulling action is to roll out the sand and bond like a rolling pin rolls out dough. The pressure of the roller is important. If there is too much pressure, the roller cuts completely through the sand to the base of the machine. If there is not enough pressure, the roller goes over the top of the sand mixture without shearing the mass. The mass must be compressed, sheared into sections and re-mixed to properly distribute the water and bond. The lugs and ploughs are positioned to fragment, gather and redirect the sand back under the muller wheels.

Some mullers are adjustable to account for changes in the consistency of the sand. The muller is started off with lighter pressure. As the bond is built up and the sand becomes more resistant to compression and shear, the pressure is increased. The height of the mixture increases as the clay absorbs water and swells.

In order to have as consistent sand properties as possible, the muller should be as large as practical so that the largest batch is made at one time. It is therefore best to have a larger slower turning muller than a small faster turning muller of the same horsepower. A few large batches are better than several small batches.

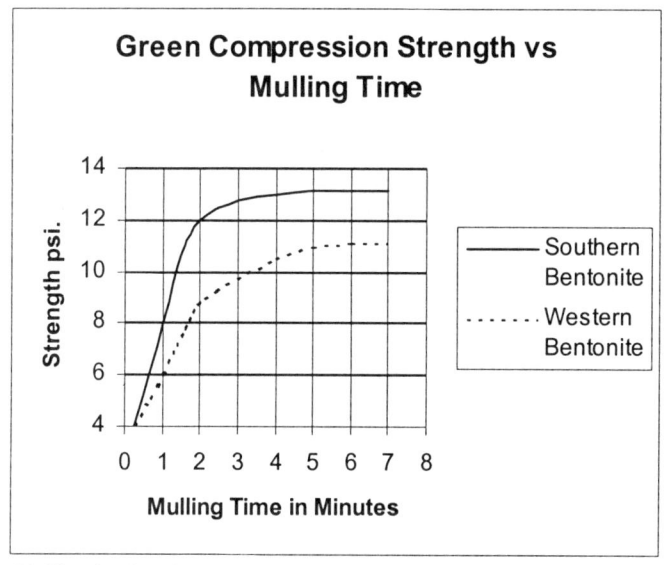

Green Compression Strength vs Mulling Time

Mulling time based on AFS #60 sand with 5% bentonite and 2.5% water by weight.

Typical Mulling Sequence

1. Add all the base sand and any wood flour or cellulose that may be used
2. Mull for at least 30 seconds up to 2 minutes.
3. While continuing to mull, quickly add all the temper water. *(note that some foundries prefer to add the clay and water as a slurry) Continue to mull
4. When the water is well distributed, add the clay and continue to mull
5. Add any seacoal or carbons
6. Mull for an additional 3 minutes, finer sands may require longer mulling times to fully distribute the bond

Many foundries have changed sand systems from green sand to air-set sand. They have no use for a muller and may have one sitting idle in a back room. Often, they can be purchased inexpensively. Below is a photo of my 5-hp muller. Many mullers use two wheels.

1940's Simpson Muller

CEMENT BINDERS:

Cement bonded molds may be made with considerable accuracy and are often used for large molds requiring more accuracy than can be attained with other molding methods. Cement is often used as a shake on facing to give a very high dry strength on the surface of a damp mold. Sometimes a pattern is returned to the mold and rapped down to set the coating. In order to prevent the cement from sticking to the pattern, the mold is dusted with finely ground charcoal. A mixture of cement and finely ground charcoal is used on detailed brass and bronze work. It is dusted on the surface of detailed molds to get a better finish and to keep lettering from washing away from plaques.

Cement bonded sand may be used anywhere the metal shrinks away from the mold wall or core. Cement bonded cores have limited use because they have such high strength. Large cores that are moved by cranes require very high strength. Cement bonded cores have the required strength and also have hard surfaces. They are used on large stationary and marine diesel engines frames. Other applications may be large molds such as lathe beds and other machine frames.

Portland cement is generally used in foundry work where cement is used as a binder. Molds and cores are made of high early strength hydraulic type cement that reaches maximum strength in 3 days as opposed to regular Portland cement that requires 28 days to cure. Lower temperatures slow the curing process and at $40°$ it stops. Proper temperature is needed for the cement to cure.

Cement bonded sand uses 8 to 12 weight percent cement and 4 to 6 percent water. Sometimes molasses water is used to accelerate the setting time. The mix should just furnish enough cement to coat the sand grains. One weight percent wood flour is added for core mixes. Patterns should be oiled for parting. Cement bonded sand should be tamped rather than rammed as ramming causes the sand to move about the mold rather than compact. Allow the mold to harden or set up before the pattern is withdrawn, however if left overnight, the pattern may become permanently set in the cement. The mold should cure for 72 hours. After curing, the water content of the molds may be reduced below 1 percent if the molds are oven baked.

Cement lowers the fusing point of sand; therefore cement-bonded molds must be blackened, usually on the day after it is made. At least two coats and sometimes three are used to help peel the mold from the casting. Coarse sand, 40 grain fineness is often used very large castings and must have several coats of blacking to keep the metal from penetrating between the grains and permanently locking them onto the casting.

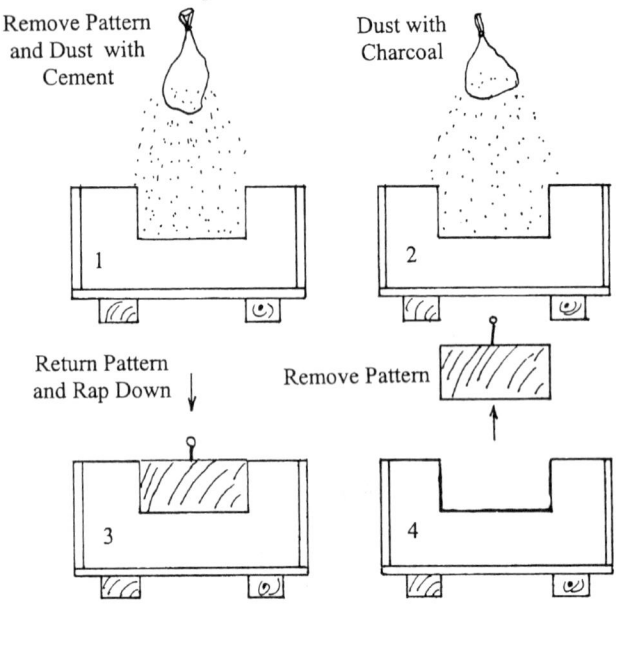

Applying Cement Facing for Ornate Work

Bronze pressure castings may be made with cement bonded sand. Cement molds release very little gas and chill the metal faster creating a thicker skin that is helpful on bronze pressure castings that may leak or sweat lead as they solidify.

Cement facing is used for copper based alloys and cast iron when the mold surface has fine detail, lettering or ornate designs as found on plaques and antique reproductions. A layer of fine charcoal keeps the cement from sticking to the pattern.

DEVELOPMENT OF OIL BONDED SAND:

Many of the properties of conventional molding sand depend greatly upon the moisture content. During working, and storing, the moisture content of the molding sand can vary due to evaporation. Slight changes in moisture content can cause large changes in the sand's properties leading to various casting defects. It is also more difficult to maintain the proper water ratio in hot climates.

In order to vent the large amount of steam produced when molten metal contacts the sand, conventional molding sands must be permeable or porous. The permeability is related to the grain size, with larger grains being more porous. This limits the fineness of the sand that may be used for a particular casting. Smoothness of surface finish is also related to grain size with smaller grains producing smoother surfaces.

Oil, because it has a higher boiling point than water and evolved less gas, was believed to be a good substitute for water. Because of its much lower heat transfer rate, oil also causes less chilling of the molten metal.

Clays-oil mixtures are impractical as bonding agents because they develop very low strength. The National Lead Company developed a bonding agent effective in oil that consisted of a mixture of metal oxides and bentone. They were mulled into the sand with oil and a catalyst of ethyl or wood alcohol. This mixture was tested at the Armour Research Foundation of Chicago during 1954 and 1955 and was first introduced to the casting industry in 1956.

Bentone is produced by reacting sodium bentonite with organic ammonium or onium salt to produce an organic onium bentonite plus salt.

Aluminum and brass castings are produced with a good surface finish using bentone-bonded sand. The sand may be used several times without rebonding; often riddling between use is sufficient. Because of the lower heat transfer rates, the ratio of sand in the mold to metal poured becomes more important than with conventional sands. Metal temperature and time before shakeout are also important. As long a time as is practical between pouring and shakeout allows more of the oil vapors to condense in the mold. Shaking out too soon may cause the oil vapors to ignite causing a flash fire. Shakeout is usually ventilated to remove the smoke and fog that is produced.

PETRO BOND

Petro Bond is a waterless binder that is used with lower pouring temperature nonferrous alloys such as aluminum and bronze. Petro Bond binder was developed by the Bentonite Corporation as a "sand casting" alternative to low volume diecasting. Petro bonded sands are very fine. Molds have a green strength 12 psi. They give an excellent surface finish and are ideal for casting automotive trim parts that are going to be plated. Petro Bond sand is also used in hot climates where water based sands quickly dry out.

Petro Bond sand is a mulled mixture of sand, oil, Petro Bond, and a catalyst. Since there is no water used, less gas is generated. This allows the use of very fine sands with lower permeability. Silica sand and olivine sands with a grain fineness of 120 to 180 may be used for aluminum and copper based alloys. Coarser grades are used for larger castings and higher temperatures. Because clays absorb water, which is detrimental to the bond, the sand must be clay free.

Petro Bond oil is made without inhibitors and for the maximum properties, is recommended over lubricating oil. Some foundrymen use 40 or 50 wt non-detergent motor oil.

Typical mix ratios by weight:

100 pounds dried sand

5 pounds Petro Bond
2 pounds Petro Bond Oil
1 ounce P-1 catalyst*

*Petro Bond II does not require a catalyst

Mulling Procedure for Standard Petro Bond:

Mull 100 pounds of sand and 5 pounds dry Petro Bond for 30 seconds
Add oil and mull for 3 minutes
Add catalyst and mull for 8 minutes

Mulling procedure for Petro Bond II

Add 100 pounds of sand and 5 pounds of Petro Bond II
Mull for 30 seconds
Add 2 pounds Petro Bond Oil and mull for 6 minutes

Petro Bond must be mulled. Hand mixing does not work. Proper mulling time is essential for green strength. The minimum green strength for properly mulled sand is 8.5 psi. and the maximum is 12 psi. In order to get the maximum properties, mulling times may be longer than those specified above. Green strength vs. mulling time is shown in the graph below (sand grain fineness 140).

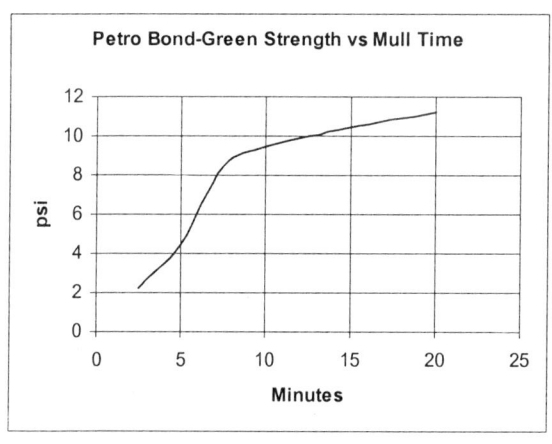

Petro Bond-Green Strength vs Mull Time

Because there is less gas generated with Petro Bond binders, the molds may be rammed harder. The best casting finishes come from very smooth hard rammed molds. Never use liquid parting compounds with petro bond sands, use parting dust or dry Petro Bond power.

Because the heat transfer to the oil bonded sand is less than that of a clay-water bonded sand there is less chilling effect. The molds are often poured at lower temperatures than with conventionally bonded sand. Because the cooling rate is much lower for Petro Bond sands, the grain structure is coarser and mechanical strengths are lower than with diecasting or conventional greensand binders. Higher mechanical properties, strength and hardness, are attained by proper heat treatment of the completed casting. Molds formed with Petro Bond sands should be left much longer to cool than similar green sand molds.

Upon shake out, there will be a small layer of burned sand on the casting. This sand should be discarded, not returned to the molding sand. Petro Bond sand may be reused after riddling. It requires little effort to maintain Petro Bond sand.

Diagnosing problems when using Petro Bond sand:

Low Green Strength: With proper mulling, the minimum green strength should be 8.5 psi.

1. **Improper mulling**. The muller must be clean and dry. The wheels for the muller should be lowered to the pan. Smaller mullers may require longer mulling times.
2. **Not enough Petro Bond**. Make sure that the dry contents are well mixed before the oil is added. Add the oil slowly and be sure none leaks out of the muller.
3. **Moisture in the sand**. The moisture content must be lower than 0.25%.
4. Improper oil used. If the finished mix has a glossy appearance, the wrong oil was probably used.

Poor casting finish:

1. Mold not rammed hard enough.
2. Wrong parting agent. Use a dry powder or dry Petro Bond powder.

Cutting or washing into the mold:

1. Soft ramming.
2. Improper pouring. Petro bond makes a very smooth mold surface. Metal flows quickly over this surface and the sprue may need to be choked to reduce the pouring rate. Smaller gating systems may also help reduce turbulence.
3. Improper oil used.

Poor pattern reproduction:

1. Soft rammed molds.

Out of dimension castings:

1. Cooling time is too short. Petro Bond sand conducts heat slowly. Castings take much longer to cool than with usual green sand molds. Castings may warp if they are taken out too soon.

Gas Defects:

1. Too much oil in the mix. Dilute the mixture with fresh sand and dry Petro Bond powder. Mull it to proper green strength.
2. Too much catalyst in the mix. If there is too much catalyst, then there is an increase in oil absorption. This gives the sand a dry feel. Do not add more oil to the sand. Two ounces is the maximum for 5 pounds of dry binder. Gas defects occur when oil is over 10%.

FURAN BINDERS:

Furan no-bake binders are used for both molds and cores. Furan sands have excellent flowability and develop high strengths allowing you to mold parts that would otherwise be too difficult or impossible to cast in green sand. Furan sand may also be used as facing sand. In order to reduce cost of larger molds, patterns may be faced with furan sand and backed with green sand after the mix cures.

Furan binders are resins that harden upon reaction with a catalyst such as phosphoric acid. The resin may be proportioned to set up quickly or over several hours. The rate of reaction is also temperature sensitive. Warm sand and temperatures will increase the speed of cure. For every 18° F, the speed of cure doubles. The tensile strength and hardness of the mold are similarly adjusted by proportioning the components of the sand system.

Furan binders are available with different levels of nitrogen and water. Lower levels of nitrogen and water help reduce binder related gas defects. The cost of the binder increases as the nitrogen and water content decreases.

Wear rubber gloves when working with furan binders.

The primary component of furan resins is furfuryl alcohol, which is a solvent to shellac and some paints. Bare wood patterns and coreboxes are also used with furan binders.

Rubber gloves should be worn when mixing and handling furan sands to prevent staining of the hands. The dark furan stains are difficult if not impossible to remove and may last a week or more.

Mixing procedure is to combine the sand and catalyst and mix for at least two minutes. Add the binder and mix until uniform. The batch should be used as soon as possible. If added in the reverse order, the catalyst quickly sets the binder as localized hard spots and prevents further mixing. **Furan binder and catalyst should never be mixed without sand. A violent reaction and possible explosion will occur.**

Southern Foundry Resins and Ashland Chemical are two of many commercial suppliers of furan binders. The properties of a sample batch of Ashland's Chem-rez-284 and Catalyst 2009 are seen below*.

Binder	CHEM-REZ 284 @ 1.2% weight of sand
Catalyst	Catalyst 2009 @ 35% weight of binder
Working Time in Minutes	4.5
Strip Time (from mold) in Minutes	8.3
Tensile Strength in psi at 1 hour	141
Tensile Strength in psi at 3 hours	279

*The sand is machine mixed at 72°F

Normal catalyst levels are in the 25 to 45% range; however it is not uncommon to use more or less depending upon the properties required. Ashland recommends use of sulfonic acid rather than phosphoric acid for their products.
Storage and Handling: Furan binders should be stored between 40°F and 90°F. Catalysts are strong acids and protective clothing should be worn. **Mixing catalyst and binder in the absence of sand may cause an explosion.**

CORES:

Generally, cores form the inside of a casting. Cores form all the inside surfaces of an engine block, a piston and an iron-plumbing elbow. However, cores are not limited to the inside of castings. Cores may form any part that is not formed by the pattern in a sand mold. Generally cores are used where green sand would not have the strength to hold up under the hot flowing metal. The fins on an air-cooled cylinder are made in core sand.

An Air Cooled Cylinder Cast in a Baked Core Sand Mold

There are two types of cores, green sand and dry cores. Green sand is self explanatory, the sand is still moist in this type of core. Dry sand cores may harden by standing, or after some chemical treatment such as gassing sodium silicate with CO_2. Many cores are baked to reach the proper hardness. Cores that use binders such as molasses, wheat flour, or linseed oil are all hardened by baking.

Green sand cores left by the pattern are the cheapest type of core. They require no extra mixing, a core oven, drier plate or any of the other equipment associated with coremaking. There are many applications where green sand cores are not practical. In such cases, dry sand cores are used. The majority of cores used in the foundry are dry sand.

Machined Cylinder from the Baked Sand Mold Above

Cores must be properly located relative to the mold. Patterns for parts that use cores have extensions on them called core prints. The prints leave a void in the mold so that extensions on the core may be placed into the "core prints" formed in the mold. The headstock casting below has core prints for the center core and one for the bottom. Note the voids in the casting where the core prints were.

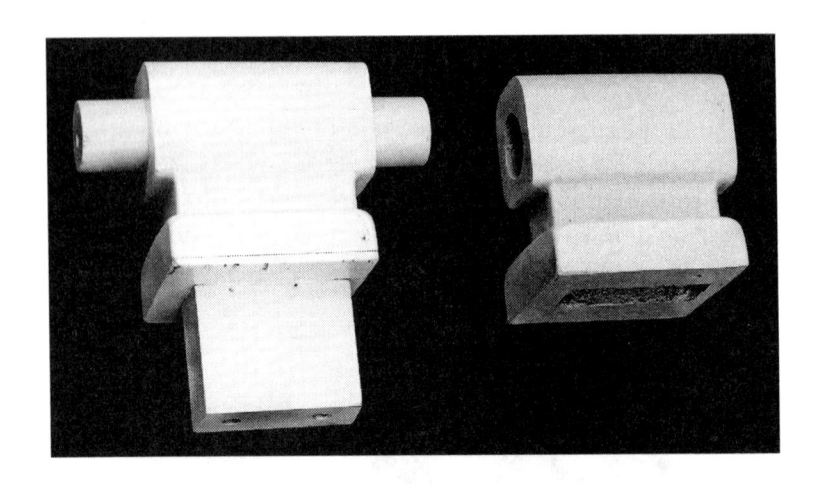

Headstock Pattern and Casting

Notice the large core print for the hanging core in the base of the casting

Sand Cast Piston Blank and Core

The inside of a piston is formed by a core. The core may be composed of many metal parts or it may be made of sand as shown above. Both the piston blank and core have been sectioned so that you may see the matching contours.

MAKING CORES:

Cores are made in a **core box,** which is a wood, or metal box that has a cavity in the shape of the core. Core boxes may be very simple in design or they may be composed of several pieces to form an intricate core. The core box is rammed with core sand and flipped over on a flat metal sheet called a core plate. The core box is rapped or vibrated then lifted leaving the core on the plate to dry. If the core does not have a flat surface, a specially shaped core plate

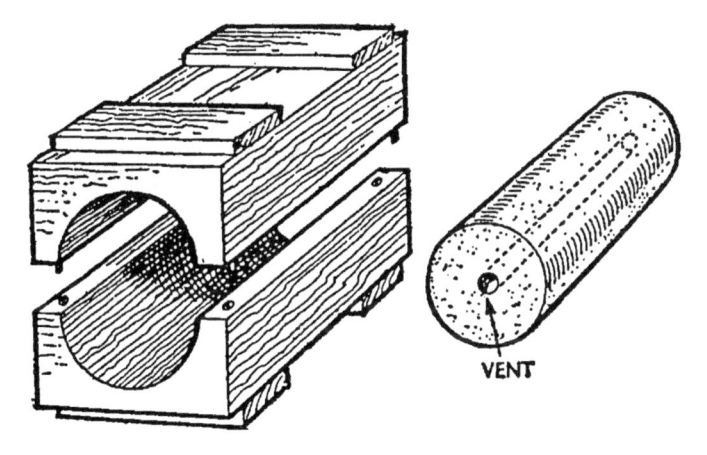

A Cylindrical Core and Core Box

> Note that a half round core box is often used to make a flat surface for baking. The two halves are glued together after baking to make a cylindrical core. This eliminates the need for a round core drier.

that conforms to the core called a core dryer is used. A round core would sag out of shape on a flat plate so a rounded core dryer is used to maintain the proper shape of the core. Some cores are made in a core box where one half of the box acts as the drier. The two halves are clamped together and the core sand is rammed in. The top half is removed and the core is left in the bottom half to be dried. This produces seamless cores. Sometimes the cores may stick to such driers. Coating the drier with a thin film of oil or no stick baking spray and then some parting compound may relieve the sticking problem.

Engine Cylinder & Crankcase Core Box with Drier

Vent rods and reinforcing rods are shown in place. The drier is placed over the core and a flat plate is placed over box and drier. The assembly is turned over, the vent rod is removed and the core box is rapped and lifted off the core. The dried core is shown below.

Core plates and dryers must be flat or straight not to distort the cores. Small cores baked on a cookie sheet are often distorted because the sheet will warp when it gets hot. Small core plates can be made of steel plate. They are often made of aluminum sheet. They may also be made of cast aluminum and machined to final dimension. Cast iron core plates are used in large production runs. Core plates should be plane within .015 for quality work. Straighter is better.

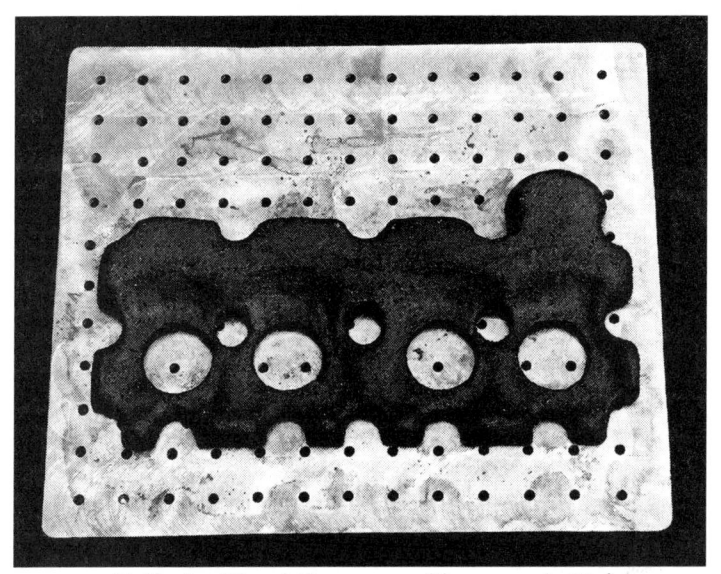

Cylinder Head Core on a Shop-made Drier

This drier is cast in aluminum, machined flat and drilled. There are reinforcing ribs on the bottom side of the plate. The drier pattern is mounted on a match-plate. I use the unmachined castings for bottom boards and squeezer boards.

VENTS and VENTING CORES:

When molten metal comes in contact with a core, gasses are formed from the burning binder that must be vented. Core sands are permeable and may be all that is needed to vent the gas away from the casting. The more of the core that is surrounded by metal, the less open area for venting gasses to the outside of the casting. More venting is needed in this situation. Ramming up a vent rod or pushing a rod through a core works to form most vents. Where a straight rod can not be used, vent wax is used. It comes in boxes or on a spool from 1/32 to 5/8 inch in diameter. As the core bakes, the wax is melted leaving an opening for the gasses to escape to the outside of the casting. Sometimes a smooth cotton or nylon string or rope is threaded through a hole in the core box and rammed up with the core. It is

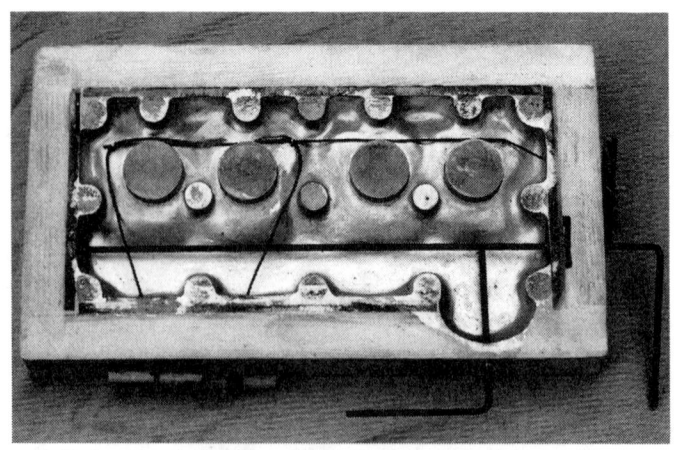

Cylinder Head Core Box With Vent Rods and String Vents

pulled out through the hole to leave a vent. Nylon builder's string works well for small vents. Sometimes vents are formed with several rods that touch each other at right angles. When the vent rods are removed the holes on the surfaces that are exposed to metal are plugged to prevent metal from entering the vent and plugging it during casting.

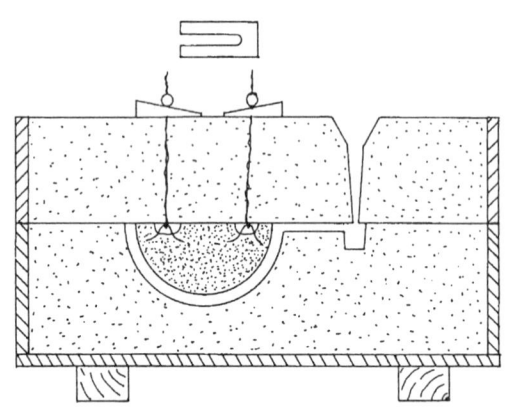

Core suspended in the cope by wires

Rods or wires often reinforce cores so that they will withstand the buoyant force caused by being submerged in molten metal. Wires or rods may also be added for lifting and securing the core in the mold.

166

Re-bar tie wire works well for small cores. Re-bar and threaded rod works well for larger cores. Small reproduction cannon barrel cores may be formed over a section of pipe wrapped with burlap and coated with a core mix. The cast shell below uses a similar arbor to hold the core. Note the vent holes drilled into the pipe.

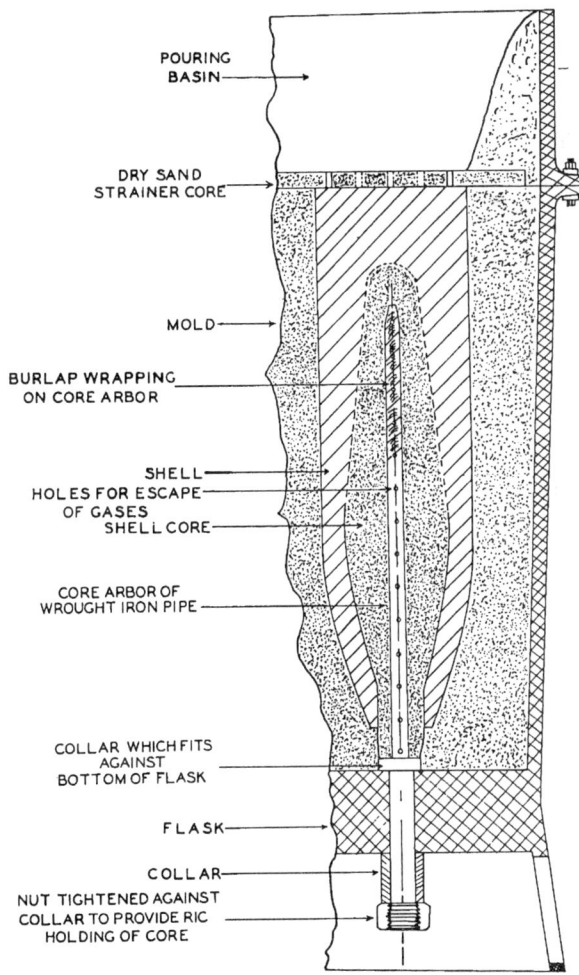

A Cast Iron Bomb Shell Core

Considerable skill is needed to remove a deep core box from a core without breaking it. In such difficult situations, a "roll over core machine" may be used. The machine bolts to the core box. The core box is rammed. The dryer is set into position and clamped in place. The whole assembly is rolled over and the core box is removed by the lowering the core plate on the machine. There are several types of roll over machines.

Roll Over Machine

In large operations, core sand is blown into the mold with compressed air. Core sand is blown in through a small hole or holes in the core box. The box vents must be carefully positioned so that the sand fills all the corners of the mold and they are often found by trial and error. Core sand must have high flowability to properly fill the core boxes.

PROPERTIES OF CORES:

The three most important properties of a core are **1**. Its ability to stand up to the flow of molten metal without having sand grains wash from its surface. **2**. The ability of the binder to burn out soon after solidification has started so that the core may collapse. This allows the casting to contract with out cracking as it is cooling. It also allows the core to be easily removed. **3**. The core must generate a minimum amount of gas upon burning out. This minimizes gas bubbles in the casting.

CORE BINDERS:

Although there are many materials available to the commercial foundryman, our discussion of core binders is principally limited to simple and readily available binders such as linseed oil, molasses, wheat flour, and wallpaper paste. Each type of binder has properties that make it useful and they may be combined to get the best properties from each. Linseed oil has very little green strength and it is very difficult to get a core out of the box without deforming it when using linseed oil alone. Wheat flour and molasses water gives good green strength that allows you to get a core out of the mold. They have moderate baked strength. Both types of binders may be combined to get the best of all the properties.

The burning of the binders in a core generates gas. Therefore, *whichever binders you use, you should use as little as possible to get the desired results.*

Molasses is a by-product of the sugar refining industry and is produced from both cane and beet sugar. "Blackstrap molasses" is produced from cane sugar and is available in grocery stores. A solution of 10% molasses and 90% water is often used in core work. The solution may be used to temper the core sand and /or it may also be sprayed on a core or mold surface to provide a hard crust. Because molasses is composed of mostly sugars, it is affected by the moisture content of the sand mixture with higher moisture promoting a harder shell. This is because the excess moisture carries the molasses to the surface as it evaporates.

Molasses rapidly absorbs water from the air. Molasses cores should not be left standing in open air or be allowed to sit in green sand molds but instead be taken from the oven, placed in the mold and the mold poured immediately. Sand tempered with molasses has little green strength and is usually mixed with a cereal type binder like wheat flour.

Such cores bake fairly hard and may be easily handled. Molasses and wheat flour binders burn out well in small iron castings and are easily removed by pouring, like dry sand, from the casting.

Cereal Binders include wheat four, corn flour and dextrin. Wallpaper paste, used for hanging wallpaper, is a much stronger binder than wheat flour; therefore, cores require less binder, are stronger, easier to work and produce less gas than wheat flour cores. When hand mixing wheat flour binders, ratios as low as 1:12 by volume can be used, however the green strength may be low. Adding more wheat flour up to the point where the mixture becomes doughy forms higher green strengths. Generally 1 part flour to 6 or 8 parts sand by volume makes cores strong enough to work for hand mixed home foundry work. Increasing amounts of binder creates larger amounts of gas.

Corn flour is made when corn meal is heated in water and then run across hot rollers to gelatinize and flake it into grits. The flakes are then ground and screened. Dextrin is powerful binder made from cornstarch. It is usually white or yellow in color and dissolves completely in water. It migrates to the surface of the mold like molasses giving a hard surface and sharp edges. Dextrin is made by boiling cornstarch with a mixture of dilute nitric and hydrochloric acids for a given amount of time.

Cereal binders are used for green strength, or to keep the core from falling apart before it is baked. For maximum baked strength when using oil binders, the percentage should be less than 3 times the percentage of the core oil in the mixture. An amount or ½% to 2% is used to give about 1 to 2.5 psi green strength. Water must be added to develop a green bond. The water addition is usually from 2 to 5 times the percentage of cereal in the sand mix. Lower amounts of water give lower baked strengths along with lower surface and edge hardness. The cores scratch more easily.

Linseed oil produces harder cores than molasses. Linseed oil is usually mixed with from 40 to 80 parts sand to 1 part oil, using as little oil as needed to get the required results. These ratios are used with mechanical mixing. You may or may not be able to use such low ratios. When mixing by hand, use ratios up to 25 to 1. Kerosene and water are often added in small amounts to linseed bonded sand to keep the mix from sticking to the core box. Linseed oil binders have almost no green strength due to the lubricating action of the oil on the sand. The green strength

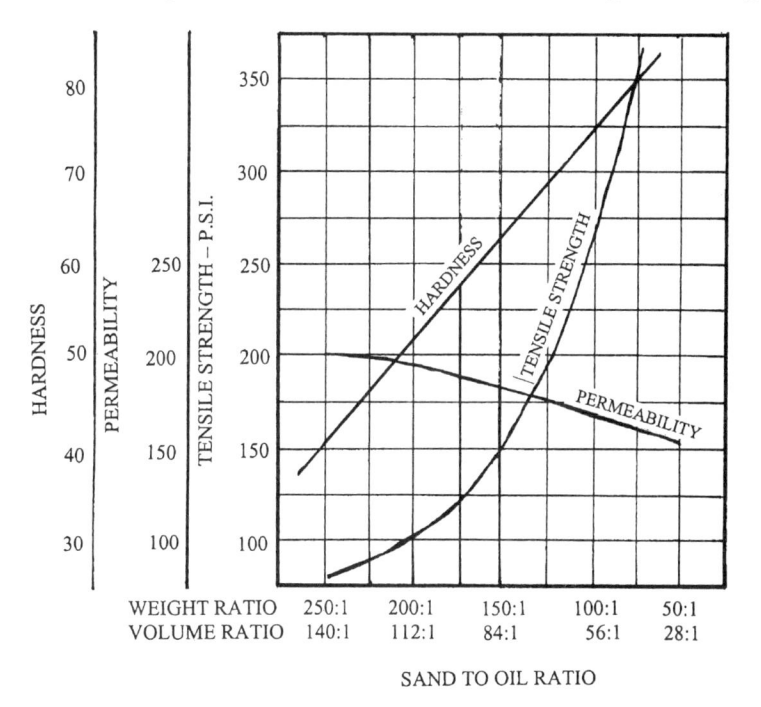

WEIGHT RATIO	250:1	200:1	150:1	100:1	50:1
VOLUME RATIO	140:1	112:1	84:1	56:1	28:1

SAND TO OIL RATIO

may be improved by adding fine sand such as bank sand or a cereal to the core mix.

When cereal is used with an oil binder, both the order of ingredient addition and the mixing time are important. When the ingredients are mixed in the order *cereal-water-oil*, the highest green and baked strength is found.

The poorest order of addition is oil-cereal-water. When the cereal and water are mulled together, they form a paste and both the green and dry strengths are high. When the cereal and water are separated during the mixing cycle or when the sand grains are first wet with oil, the strengths are low.

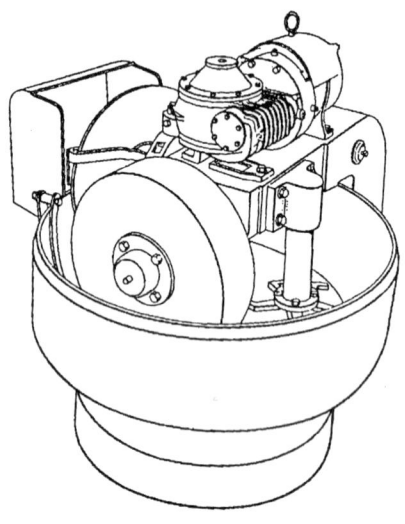

Mechanical Core Mixer

Linseed oil dries by oxidation with the reaction being similar to drying paint. The linseed oil takes up oxygen and forms larger molecules that polymerize to form hard films and hold the sand grains together. Linseed oil takes up oxygen slowly at room temperature and must be baked to speed up the process. There is a definite baking time and temperature to achieve maximum hardness. Using japan dryer, available at paint stores may speed up the process. Sometimes 0.1% ammonium nitrate or 0.03% sodium nitrate is added to speed drying. Driers can reduce the baking time by 20 to 80%. Sodium and ammonium nitrates should be purchased from a commercial source because they can make it less hazardous when baked in an oven. The graph on the next page illustrates the relationship between time, temperature and tensile strength for small cores.

BAKING CORES:
Large cores are more difficult to bake for a number of

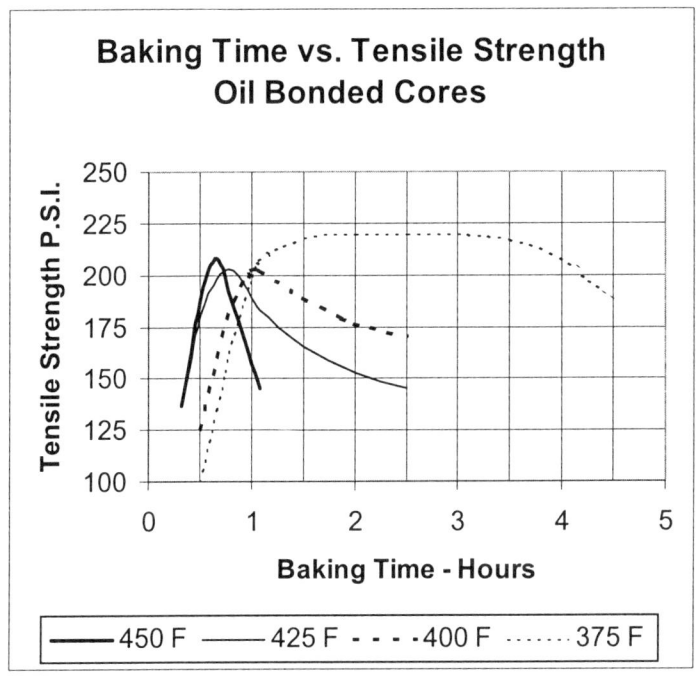

Baking Time vs. Tensile Strength Oil Bonded Cores

(Graph: Tensile Strength P.S.I. vs. Baking Time - Hours, showing curves for 450 F, 425 F, 400 F, and 375 F)

reasons. First, the outer surface will reach the baking temperature long before the center of the core. The outer surface will reach maximum strength and start to decline in hardness before the center reaches maximum strength. In such cases lower temperature and longer bake times are used. Large cores may also be hollowed out and filled with dry sand after baking.

Permeability of the sand is also a factor in core baking, because oxygen must reach the inside of the core for oxidation to take place. Cores are baked on perforated plates or dryers to reduce the variation of hardness from the outer surface to the center of the core.

BAKING TEMPERATURE and GAS CONTENT:

When in contact with molten metal, cores produce gas. This gas may blow holes in thin pressure tight castings such as cylinder liners and cylinder heads. Gas liberation may be reduced with some loss in tensile strength by baking at a higher temperature. By baking at a high temperature and slightly over-baking, a 25% reduction in gas liberation and at a cost of 25% reduction in tensile strength. Cylinder head and liner cores are often slightly over-baked.

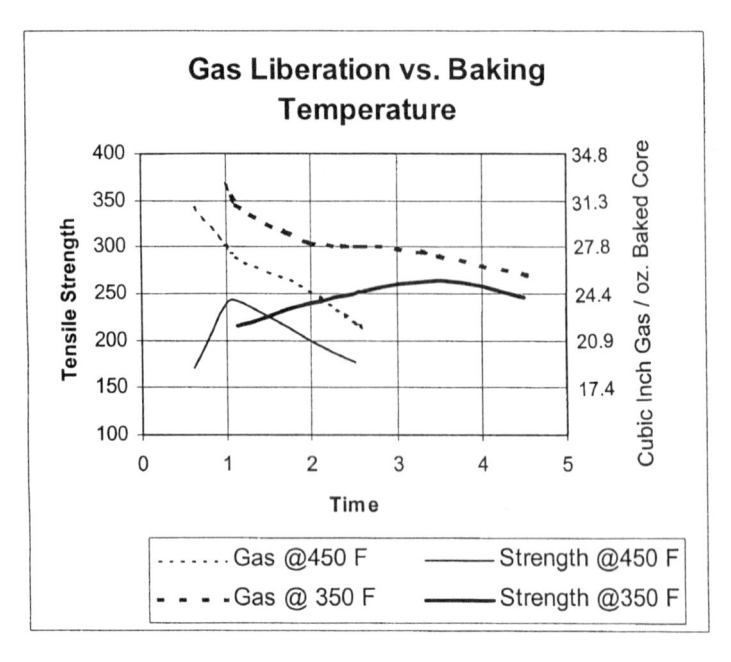

Gas Liberation vs. Baking Temperature

Time	Tensile Strength	Cubic Inch Gas / oz. Baked Core
0	100	
1		34.8
2		31.3
3		27.8
4		24.4
5		20.9
		17.4

······ Gas @450 F   ——— Strength @450 F
- - - - Gas @ 350 F   ——— Strength @350 F

AIR SUPPLY FOR BAKING CORES:

The properties of cores are related to the baking temperature, time, humidity, and amount of oxygen present in the core oven.

The amount of oxygen required to thoroughly bake a core is relative to amount of air that must be supplied to a core oven. Linseed oil requires 73 percent of its weight in oxygen to properly bake. At sea level and 400°F, air weighs approximately .0461 pounds per cubic foot. Oxygen, by weight, equals .302 pounds per pound of air. Under perfect conditions, assuming all the oxygen in the air combined with the linseed oil, 1 pound of binder would require 52.4 cubic feet of air at 400°F. In real life, the oxygen would not all combine with the oil, and if a gas burner is used to heat the oven, the amount of available oxygen would be much lower so the amount of air required would be much higher. This example illustrates the need for a certain amount of ventilation in the core oven.

The properties of cores may often be inconsistent. Some days they may be good, hard and burn out well with little gas and smoke. On other days they may be under-baked and smoky or over-baked and brittle. Atmospheric humidity plays a significant role in the properties of cores.

To discuss the role of humidity on core baking we need to define a few terms:

Saturated air: Air that contains the maximum amount of water vapor it can hold at a given temperature. This is 100% relative humidity.

Relative Humidity: The amount of moisture in the air relative to the maximum amount it could hold at that temperature (saturated air). 75% relative humidity means that the air is 75% saturated. At 75 degrees F and 75% relative humidity, air contains .001 pounds of water per cubic foot.

The weight of air: is .075 pounds per cubic foot.

At 400° F, saturated air contains .0857 pounds of water per cubic foot. In practice, 60% is the approximate maximum relative humidity useful for drying cores.

$$.6 \times .0857 \text{ pounds} / \text{foot}^3 = .0514 \text{ pounds} / \text{foot}^3$$

This leaves a useful capacity of approximately .0343 pounds of water per cubic foot of air in the furnace.

The combustion of one pound of propane produces 1.63 pounds of water vapor.

To illustrate the need for air circulation due to humidity, an example core oven is presented as follows:

A small foundry runs a core oven with a capacity of 250 pounds of cores per hour. A propane burner that consumes 3.4 pounds of propane per hour heats the oven to 400 degrees F. The input air is at 75 degrees F and 75% relative humidity. The core sand contains 6% moisture as it enters the core oven.

The amount of water to be removed from the furnace may be calculated as follows:

The amount of water in the cores is .06 x 250 pounds = 15 pounds.

The amount of water formed by the burning of the propane:

3.40 lbs. of propane x 1.63 lbs. of water / lb. propane = 5.54 lbs.

5.54 lbs. of water are formed by the combustion of 3.4 pounds of propane.

The total amount of water that must be removed from the core over per hour is:

$$\begin{array}{r}15.00 \text{ lbs.} \\ + 5.54 \text{ lbs.} \\ \hline 20.54 \text{ lbs. of water}\end{array}$$

The amount of water to be removed from the oven per hour is 20.54 pounds.

The approximate number of cubic feet of dry air needed to dry the cores are estimated as follows:

(water to be evaporated) / the capacity of air at 400° F

20.54 pounds / .0343 pounds per cubic foot = 599 cubic feet of air before adding the relative humidity of the input air. An additional .303 pounds[1] of water are in the input air thus increasing the required minimum airflow through the furnace. Most foundrymen would not perform these calculations unless it was critical for the operation. This example is to show that *air circulation is important for proper core baking* and that the minimum air circulation may be calculated.

[1] $V_2 T_1 / T_2 = V_1$, (600 cubic feet)(75 + 460) degrees Rankin / (600 + 460) degrees Rankin = 303 cubic feet, 303 x .001 pounds / cubic foot = .303 pounds

HUMIDITY AND CORE STRENGTH:

Oven humidity affects the tensile strength of a core with lower humidity producing stronger cores as show in the graph below.

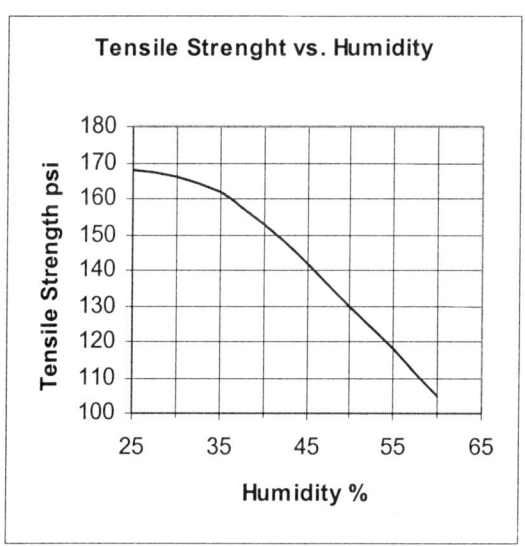

Tensile Strenght vs. Humidity

(Tensile Strength psi vs. Humidity %)

With oil, molasses and cereal binders, humidity continues to be a factor after baking because high humidity decreases

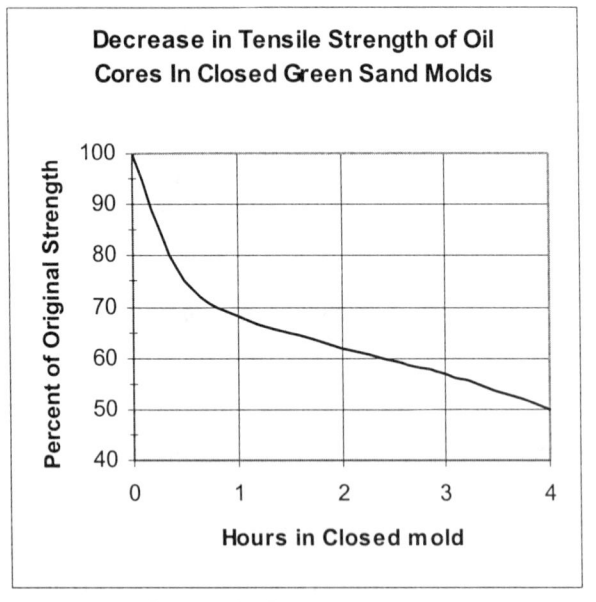

Decrease in Tensile Strength of Oil Cores In Closed Green Sand Molds

[Graph: Percent of Original Strength (y-axis, 40 to 100) vs. Hours in Closed mold (x-axis, 0 to 4). Curve drops rapidly from 100% to about 70% in the first hour, then gradually to about 50% at 4 hours.]

core strength. Since air in a closed greensand mold is saturated, 100% relative humidity, the strength of cores decreases rapidly in the first hour in a closed mold. Cores standing in a greensand mold for an hour or more will loose about 50% of their strength. When maximum core strength is needed, the mold should not be closed until immediately before the metal is ready to be poured into it.

OTHER CORE MATERIALS:

Silica flour is Finely ground silica. In the ceramic industry it is called "flint flour." Silica four is usually 200 mesh or finer. It is added to increase hot strength and reduce penetration and cutting of the molten metal. It may be used in long thin cores or where the metal remains molten for a long time. It may be added from 1% to 30% by weight. Test data has shown a 5% addition of silica flour raised the hot strength of a core sand from 1 psi to 42 psi at $2500°$ F. Too much silica flour may cause cracks and hot tears because of the very high hot strength it gives a core. Silica flour increases the time of burn out and core collapse. The addition of silica flour requires additional binder.

Sawdust and wood flour are added to weaken or soften a core and increase its collapsibility in a cooling casting. Sawdust may be used in bronze castings and in malleable castings due the high shrinkage of white iron. Finely ground material is added from 1 to 10 by volume. Sawdust additions give poorer surface finish on castings because the dust burns out quickly leaving small voids between the sand grains on the surface. Commercial wood flour, ground hardwood cellulose, may be added from less than 1 up to 3 percent. A 2 percent addition by weight, will decrease the hot strength about 35 per cent.

CORE FINISHING:

Depending on the required accuracy of the core, there may be several additional finishing operations after the core

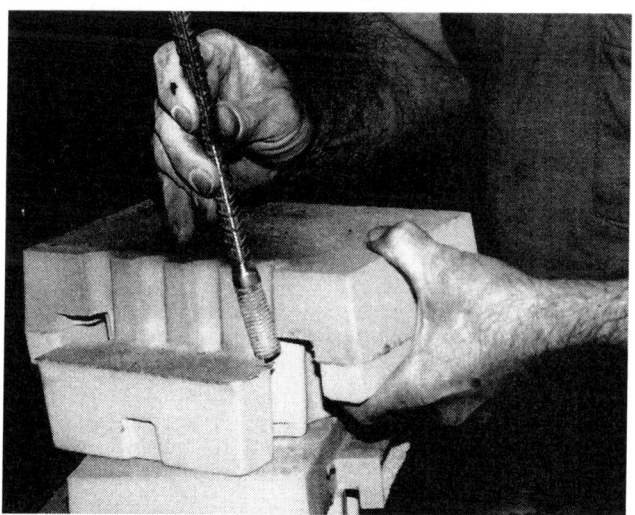

Cleaning and Filing a Core

The shop made file is a section of threaded rod welded to a re-bar handle

is baked. These finishing operations fall under three main categories, **cleaning , sizing,** and **assembly.**

Cleaning may include trimming fins or lumps from a core. Coating the core with a protective wash may fall under the cleaning category and will be discussed later. Filing, sanding with sandpaper or rubbing with a section of an old grinding wheel removes fins and lumps. Loose sand is removed by brushing. Rough spots and soft rammed places are filled with a graphite molasses-water solution mixed to a thick mud and smoothed with a finger. Graphite may also be mixed with linseed oil. Cores must be dried after filling. Use a torch or a short trip to the oven to evaporate the water. Check the core vents to be sure that they are open.

Sizing Operations: Cores may sag out of shape during baking so they are checked with gages and-or they may be filed or ground to final dimensions. Generally cores are placed on a machined plate and a template or gage is used to check the accuracy of the core.

Checking Cores with a Gage

When the core does not fit the template, it may be placed in a wooden filing fixture with protective steel strips on the filed surface. Often, the core box is intentionally made a little large to allow for filing. The baked core is placed in a fixture and filed down to the steel-topped surfaces. Engine water jacket cores may be sized like this.

Grinding an Engine Core

If many cores are needed or extreme accuracy is required, then the cores are placed in a fixture and ground to final dimension. Water jacket cores are routinely ground to accurately hold the specified metal thickness between the water jacket and the port cores. Grinding fixtures may have 4 set screws in the base to adjust the height of an individual fixture.

Assembly of cores includes pasting and bolting. In some instances molten lead is poured into cavities in the core to hold parts in place. (Remember, I am covering some old techniques in this book.) Core paste may be bought or mixed from a solution of molasses water and wheat flour until it forms a creamy consistency. Cores are pasted together and the joint is filled with a daubing mix described above.

Leaded cores are pasted together to prevent the lead from running out. The pasted joints are dried to prevent the lead from bubbling or splashing out from contact with a wet surface. Leaded cores are very strong and can not be disassembled.

Making Leaded Cores

Bolted cores: Large cores may be bolted together. Bolting is often used where pasted cores are not strong enough to resist breaking apart or floating in the rising metal. The bolts may extend through the core print and be bolted through the mold. Cores are bolted together through a hole formed in the cores by the core box. The ends of the bolt fit into recessed bosses that are filled with a plug core or are mudded over with the daubing mix.

Core Jigs and Fixtures: Core assemblies that are held to close tolerances are often assembled and set into the mold by a fixture or jig. The jig may be simple as shown setting the cores below or it may be a complex jig like those used in assembling and setting engine cores.

Setting Pump Cores Using a Setting Jig

Packard V-12 Cores

The Packard cores are assembled in a precision jig to hold the close tolerances required for an engine block casting. This jig is also used to set the core assembly into the mold.

Core Assembly for a Packard V-12 Engine

Using a jig to assemble and set the cores for a Mack 6 cylinder diesel engine.

CORE BUOYANCY:

Sand cores float when surrounded by molten metal. They float up with a force greater than their own weight. Cores are buoyed up with a force equal to the weight of the metal they displace less the weight of the core. Assume that a 1-inch square cube of core sand weighs .057 pounds and is pushed down into liquid iron until it is just submerged. Iron weighs about .26 pounds per cubic inch. The force needed to keep the core submerged is:

Buoyant force = the weight of metal – the weight of core

.26 pounds - .057 pounds = .203 pounds

If a core is 1 cubic foot it would weigh about 100 pounds and would require 350 pounds to keep it submerged. As you can see the buoyant forces on a core can become considerable.

To estimate the buoyant forces on a core for a specific metal, multiply the weight of the core by the ratio below.

Material:	Ratio
Aluminum	.66
Brass	4.25
Copper	4.50
Iron	3.50

Because cores have print areas that are not surrounded by metal the actual force will be a little less that calculated from the above table. A more accurate figure may be calculated by using the actual number cubic inches of submerged core volume. Assuming that a core weighs .057 pounds per cubic inch and then multiplying by the above ratios.

Find the force on a 3-inch diameter cylinder core that is 10-inches long and submerged in iron:

Volume of the core is: $(\pi d^2 / 4) \times$ length

$\pi = 3.14$,   d = diameter = 3 inches

$(3.14 \times (3\text{-inches})^2 \times 10\text{-inches}) / 4 = 70.685$ cubic inches

Ratio from the table above for iron = 3.5

70.685 cubic inches x .057 pounds per cubic inch x 3.5 = *14.1 pounds*

The buoyant force on the core is 14.1 pounds

ESTIMATING THE SIZE OF CORE PRINTS:

Cores are held in place by extended core prints and-or by small metal supports called **chaplets**. Two of many types of chaplets are shown to the left. If a core is too large or is not self-supporting, then chaplets are used. The general rule for green sand molds: 5 pounds per square inch is all that can be applied without distorting the mold. Since you can calculate the buoyant force on a core, the core print area is easily calculated. Using the above cylinder core calculate the minimum core area to maintain the core's position without distorting the mold.

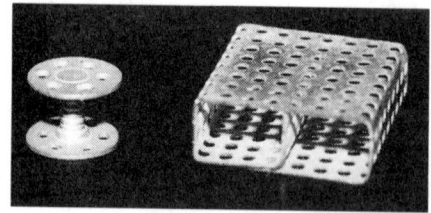

14.1 pounds ÷ 5 pounds per square inch = 2.82 inches

Now we have the area so divide by the diameter to get the length.

2.82 inches ÷ 3 inches (diameter) = .94 inches

.94 inches ÷ 2 (for each end of the core) = .47 inches

The *minimum* length for the core would be about 11 inches with ½ inch extending from each end. Of course if the core is a little longer, then you will have a little additional safety factor. Core prints may also be larger to provide more venting area.

The size and number of chaplets may be estimated the same way. Common chaplets will carry 5 psi. loading. The actual loading in psi. that steel chaplets will carry is given in the table below:

Load Safely Sustained by Chaplets in Iron

\multicolumn{3}{c	}{Double Head Chaplets}	\multicolumn{3}{c}{Stem Chaplets}
Stem Diameter	Square Head	Safe load	Round Head	Safe Load (thin metal)	Safe Load (thick metal)
1/16 inch	3/8 inch	5			
1/8 inch	1/2 inch	20	1/2 inch	20	10
3/16 inch	3/4 inch	45	3/4 inch	45	22
1/4 inch	1 inch	80	7/8 inch	80	40

The heads are sized to give the rated weight capacity on a dry core surface. Obviously green sand molds would deform under the pressure. To get around this, small bearing cores are made to distribute the pressure until the 5 psi. standard for green sand molds is reached. If a ¼ inch stem chaplet is used at the rated 80 pounds pressure then a 4-inch square (or equivalent round) bearing core is needed for the green sand mold.

80 pounds ÷ 5 pounds/square inch = 16 square inches.

$\sqrt{16}$ square inches = 4 inches.

Steel chaplets are used for iron and copper chaplets are used for brass and bronze. Aluminum chaplets are used in aluminum.

For metal to flow around the chaplet it must be **clean** and **dry**. Moisture condenses on cold chaplets placed in a damp green sand mold. It will either remain on the surface or cause rusting. Both of which will cause a faulty casting.

The chaplets must be as small as possible in order for them to properly melt into the casting *The pouring temperature of the metal must be high enough to melt and fuse the chaplets into the casting.*

Weights of Common Core Materials

Material	Pounds per Quart	Pounds per gallon	Pounds per cubic foot
Linseed oil	1.95	7.8	58.35
Wheat Flour	1.25	5	37.4
Corn Flour	0.875	3.5	26.2
Dextrine	1.4	5.6	41.9
Molasses	2.75	11	82.3
Wood flour	0.58	2.4	17.4
Sea coal	1.47	5.88	44
Carbonized cellulose	1.4	5.6	41.9
Western Bentonite	1.8	7.2	53
Southern Bentonite	1.74	7	52
Fireclay	2.4	9.6	72
Iron oxide	2.7	10.8	81
Silica sand	3.5	14	105
Silica Flour	2.5	10	90
Olivine	4.2	16.8	126
Zircon	6	24	180

Core Binders and Ingredients

Ingredient	Specific gravity	Weight per gallon	Relative hardness of cores	Bond ratio by volume	Bond ratio by weight	Moisture for machine mixing, % by volume	Mixing time min	Baking temp F	Binder destroyed F
Linseed oil	0.93	7.8	Hard	1 to 25-80	0.0312	8	5 to 7	450	600 to 700
Wheat flour**	0.6	5	Medium	1 to 12-20	.0625-.025	10	7 to 10	325	500 to 600
Corn flour*	0.45	3.5	Medium	1 to 25-50	.014-.007	10	6 to 10	325	500 to 600
Dextrin*	0.56	4.7	Medium	1 to 10-20		8	7 to 10	350	500 to 600
Molasses*	1.3	11	Frail	Mix 1:10 water	.13 water		5 to 8	250	400 to 500
Fire clay	1.2	9.6							2700
Bentonite	0.86	7.2							2700
Silica flour	1.3	10							
Silica sand	1.3	10.0-13.0							
wood flour	0.23	2							

* These binders are rarely used alone.

Binders are mixed for the best combination of properties.

** wheat flour bond may be 1 to 8 -10 for hand mixing

Automotive Core Mixtures for Gray Iron 1942

Type of casting	Sand		Bentonite			Core oil			Cereal			Water
	Pounds	Quarts	Parts	Pounds	Quarts	Parts	Pounds	Quarts	Parts	Pounds	Quarts	Parts	
1. Motor Barrel Cores		420			1.5			6			8		
2. Jacket Cores			450			8			5				
3. Automotive - unspecified			250			2			2.75			5	
4. Gear Box Cores		800						13			20		
5. Crank Case	1030			10				5.5		8			7quarts
6. Small cylinder	1000							4.5		12			8quarts
7. Transmission	1000							8			10		5quarts
8. Tappet Cores	1000			4				5.5			9		7quarts
9. Small assembly	1000							3		12			9quarts

Notes:

1,2,3,4 water is not specifed but is usually added from 2 to 5 times the percent ceral in the sand

1. 180 quarts of new silica, 240 quarts of bank sand for a total of 420 quarts, bank sand usually having 2% natural clay content

5. 730 pounds new silica, 300 pounds bank sand

7. 900 pounds new silica, 100 pounds bank sand

8. 700 pounds new silica, 300 pounds bank sand

9. 800 pounds new silica, 200 pounds bank sand

MAKING STRAINER CORES:

If a stream of molten metal from a ladle is restricted, the non-metallic material that is lighter than the metal will float out and not enter the casting. Often a strainer core is used at the bottom of the sprue or at the bottom of the pouring basin.

When liquid metal comes in contact with an oil sand core, the heat destroys the bond and the core gradually disintegrates. Since the thermal conductivity of the sand core is low, the destructive process is slow provided that the core is of sufficient thickness. Studies show that ¼-inch of material between holes is adequate for most applications. The core will survive until 1/8-inch of the material around each hole is destroyed. The core thickness also depends upon the weight of metal acting upon the core with larger diameter cores being thicker.

The size and number of holes in a core determine the rate at which metal will pass through the core as seen in the table below.

Strainer Core Dimensions

Type	O.D.	Thickness	Diameter of Holes	Number of Holes	Total area of Holes	Delivery rate in lb/sec
SC 2-3	2	3/8	1/4	8	0.39	3
SC 2 1/2	2 1/2	1/2	1/4	12	0.59	4
SC 3-7	3	1/2	5/16	12	0.92	7
SC 3-10	3	1/2	3/8	12	1.32	10
SC 3 1/2-12	3 1/2	5/8	5/16	20	1.53	12
SC 3 3/4 - 16	3 3/4	5/8	3/8	20	2.21	16
SC 4-20	4	3/7	7/16	20	3.01	20

Dimensions in inches. Flow rate determined using cast iron with a 2-inch head.

Strainer cores have a $10°$ taper on the outside diameter and a $7°$ taper on the holes.

The strainer cores shown are designed for gray iron. Selection of a core is based on flow rate. If a 30-pound, ¾ inch thick casting required 30 pounds of iron in 10 seconds, then a Strainer Core 2-3 lb./sec. (SC 2-3) should be selected.

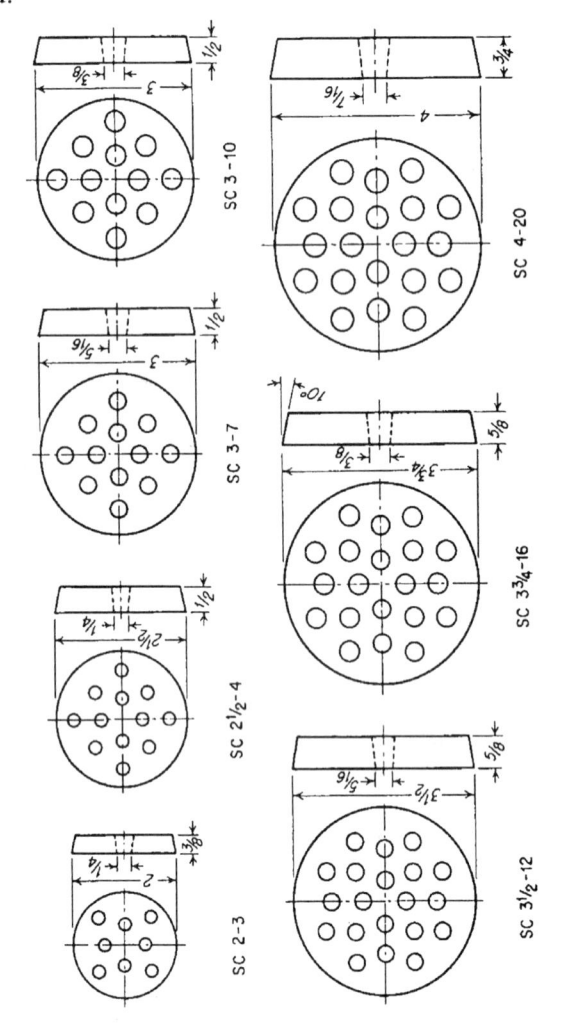

Strainer cores are commonly made in a multiple or gang box with 24 patterns. A typical mulled core mix is 95 parts sand to 4 parts oil. This mix will have little green strength.

CORE COATINGS:

Core coatings or "core washes" are used to improve the surface finish of the casting and to prevent core sand from fusing to the casting. Metal penetrates into the surface of a bare core leaving a rougher surface. Core washes fill the gaps between the sand grains providing a smooth surface and preventing metal penetration. Core washes also prevent core sand from fusing to castings by providing a refractory surface between the sand and molten metal. Core removal is easier and cleaning time is reduced. Core washes may also prevent sand from loosening from the surface of a core and washing into the mold. A core will erode when the surface collapses too fast. Refractory coatings help minimize this problem.

Metal penetration of the core caused the rough inner surface of this cylinder liner casting. It may have been prevented by using a core wash or a finer sand.

Core washes may contain carbon or other ingredients such as silica flour, olivine flour, talc, mica, alumina, or zircon. Carbon washes or "blacking" contain carbon in the form of graphite or "plumbago," carbon black, crushed coal or coke, coal tar or pitch.

Core washes are bonded with molasses water, linseed oil or clays. Alcohol can be added to the solution and burned off to dry the wash. Silicates may be used, but they do not burn out and they lower the sintering point of sand. Molasses and linseed oil burn out at 400°F and 700°F respectively. Clays are destroyed from 2000°F to 3000°F.

The dried core wash should be softer than the surface of the core otherwise it may peel off during pouring of the mold. Scabs are the result of the coating being too hard and peeling off. An excess of binder makes the wash hard and generates gas as it burns off in the mold. Use the minimum amount of binder necessary to get the wash to stick.

Cores may be painted, dipped or sprayed with the wash. Large cores may be painted with a thin wash to penetrate several layers of sand, then painted with a thick wash to build up a heavy coat, then painted with a final thin coat to smooth out the brush strokes. Be careful not to let the wash pile up in crevices or develop runs if you are spraying core washes. If the cores are washed while hot, approximately 180°F they may dry without further heating. The cores must be below 212°F when painting to prevent the wash from boiling and making a poor surface. Generally, cores are put back into the oven to dry the wash. Mold surfaces are often sprayed and dried by burning off an alcohol base or with a few light passes of a torch. You must be very careful when using a torch to dry the wash, not to burn all the binder out or over heat and crack the surface of the mold.

A **general wash formula for iron and bronze** is plumbago and molasses water mixed to the consistency of paint. Bronze cores and molds are sprayed with a thin coating of molasses water and dried. For fine work such as plaque letters this is best accomplished by spraying over the molds and letting the mist settle on the surface. This prevents the fine lettering from washing away. Another wash for brass and bronze consists of 15 parts water, 11 parts plumbago and 1 part western bentonite.

Core washes for aluminum may be of talc or plumbago and molasses water. However, finer core sands are often used to achieve a smoother surface without the use of a wash. Because the permeability of finer sands is lower, good venting is important.

Summary of Core Wash Formula

Type of Metal	Amount of Water	Refractory	Binders and other ingredients
Brass, Bronze	15 parts water	11 parts plumbago	1 part western bentonite
Steel	2.5 gallons	15 pounds silica flour	1.25 pounds western bentonite
All Metals		1 part plumbago	1 part alcohol
All Metals		1 part talc	1 part alcohol
All Metals	8 to 10 parts		1 part molasses

Bibliography

American Society for Testing and Materials, *Manual On The Use Of Thermocouples In Temperature Measurement,* (Philadelphia, Pennsylvania: American Society for Testing and Materials, 1970).

Backerud, Lennart, Guocai Chai and Jarmo Tamminen, *Solidification Characteristics of Aluminum Alloys, Volume 2, Foundry Alloys,* (Stockholm, Sweden: American Foundry Society,).

Baker, PhD., H. Dean, E.A. Ryder, M.E. and N.H. Baker, M.A., *Temperature Measurement in Engineering,* (New York, New York: John Wiley & Sons, Inc., 1953).

Black, Newton Henry, *An Introductory Course in College Physics,* 3rd ed. (New York, New York: The Macmillan Company, 1950).

Campbell, OBE, FEng, John, *Castings,* (Oxford, Auckland, Boston, Johannesburg, Melbourne, New Delhi: Butterworth Heinemann, 1991).

Capotosto, Rosario, *200 Original Shop Aids and Jigs for Woodworkers,* (New York, New York: Sterling Publishing Co., Inc., 1983).

Chastain, Stephen D., *Build an Oil-Fired Tilting Furnace,* (2002).

Dietert, Harry W., *Modern Core Practices and Theories,* (Chicago, Illinois: American Foundrymen's Association, 1942).

Gingery, David J., *The Charcoal Foundry,* (Springfield, Missouri: David J.)

Heine, Richard W., Carl R. Loper, Jr., and Philip C. Rosenthal, *Principles of Metal Casting,* (McGraw-Hill Book Company, Inc., 1967).

Jones, B.Sc., J.Inst.P., F.Inst.M.C., E.B., *Instrument Technology, Volume 1 and 2,* (London and Boston: Newnes-Butterworths, ??).

Kaiser Aluminum & Chemical Sales, Inc. *Casting Kaiser Aluminum,* 3rd ed., (Oakland, California: Kaiser Aluminum & Chemical Sales, Inc., 1974).

Marshall, Stewart, *Building Small Cupola Furnaces,* (1996).

Paschkis, M.E., E.E., D.Sc., V., *Industrial Electric Furnaces and Appliances, Volume I and II,* (New York, New York: Interscience Publishers, Inc., 1945).

Sanders, Clyde A., *Foundry Sand Practice,* 6th ed., (Skokie, Illinois: American Colloid Company, 1973).

Scott, Elmer F., *Coremaking 1940,* (Bradley, Illinois: Lindsay Publications, 1994).

U.S. Navy Department, *U.S. Navy Foundry Manual,* revised ed., (Bradley, Illinois: Lindsay Publications, Inc., 1958).

Weast, Ph.D., Robert C., *Handbook of Chemistry and Physics,* (Cleveland, Ohio: The Chemical Rubber Company, 1972).

Wellman, B. Leighton, *Technical Descriptive Geometry,* (New York,: McGraw-Hill Book Company, Inc., 1957).

Properties of Metals

Substance	Compostion	Specific gravity	Density pound / cubic inch	Mean Specific Heat 60 deg. To Melting point Btu / Pound	Heat of Fusion	Mean Specific Heat of Liquid Btu / Pound	Melting Point Degrees F	Average Pouring Temperature	Heat Content of Liquid at Pouring Temperature Btu / Pound
Aluminum	Al	2.68	0.0965	0.248	169	0.26	1215	1380	498
Brass, Red	90 Cu, 10 Zn	8.72	0.3150	0.104	86.5	0.115	1952	2250	318
Brass Yellow	67 Cu, 33 Zn	8.45	0.3050	0.105	71	0.123	1688	1950	274
Bronze, Bearing	80 Cu, 10 Pb, 10 Sn	8.91	0.3220	0.095	79.9	0.109	1832	2050	272
Bronze, Bell	78 Cu, 22 Sn	8.65	0.3125	0.1	76.3	0.119	1634	1900	265
Copper	Cu	8.96	0.3235	0.104	91	0.111	1982	2200	315
Gold	Au	19.3	0.6973	0.033	28.5	0.034	1945	2150	98
Iron	Fe	7.86	0.2841	0.165	89	0.15	2786	2912	555
Lead	Pb	11.34	0.4097	0.032	10	0.034	621	720	31
Magnesium	Mg	1.77	0.0628	0.272	83.7	0.266	1204	1380	441
Zinc	Zn	7.13	0.2575	0.107	48	0.146	786	900	142

Type K Thermocouples
Temperature in Degrees F — Reference Junction at 32F
Millivolts

°F	0	1	2	3	4	5	6	7	8	9	10
0	-0.68	-0.66	-0.64	-0.62	-0.60	-0.58	-0.56	-0.54	-0.52	-0.49	-0.47
10	-0.47	-0.45	-0.43	-0.41	-0.39	-0.37	-0.34	-0.32	-0.30	-0.28	-0.26
20	-0.26	-0.24	-0.22	-0.19	-0.17	-0.15	-0.13	-0.11	-0.09	-0.07	-0.04
30	-0.04	-0.02	**0.00**	0.02	0.04	-0.07	0.09	0.11	0.13	0.15	0.18
40	0.18	0.20	0.22	0.24	0.26	0.29	0.31	0.33	0.35	0.37	0.40
50	0.40	0.42	0.44	0.46	0.48	0.51	0.53	0.55	0.57	0.60	0.62
60	0.62	0.64	0.66	0.68	0.71	0.73	0.75	0.77	0.80	0.82	0.84
70	0.84	0.86	0.88	0.91	0.93	0.95	0.97	1.00	1.02	1.04	1.06
80	1.06	1.09	1.11	1.13	1.15	1.18	1.20	1.22	1.24	1.27	1.29
90	1.29	1.31	1.33	1.36	1.38	1.40	1.43	1.45	1.47	1.49	1.52
100	1.52	1.54	1.56	1.58	1.61	1.63	1.65	1.68	1.70	1.72	1.74
110	1.74	1.77	1.79	1.81	1.84	1.86	1.88	1.90	1.93	1.95	1.97
120	1.97	2.00	2.02	2.04	2.06	2.09	2.11	2.13	2.16	2.18	2.20
130	2.20	2.23	2.25	2.27	2.29	2.32	2.34	2.36	2.39	2.41	2.43
140	2.43	2.46	2.48	2.50	2.52	2.55	2.57	2.59	2.62	2.64	2.66
150	2.66	2.69	2.71	2.73	2.75	2.78	2.80	2.82	2.85	2.87	2.89
160	2.89	2.92	2.94	2.96	2.98	3.01	3.03	3.05	3.08	3.10	3.12
170	3.12	3.15	3.17	3.19	3.22	3.24	3.26	3.29	3.31	3.33	3.36
180	3.36	3.38	3.40	3.43	3.45	3.47	3.49	3.52	3.54	3.56	3.59
190	3.59	3.61	3.63	3.66	3.68	3.70	3.73	3.75	3.77	3.80	3.82
200	3.82	3.84	3.87	3.89	3.91	3.94	3.96	3.98	4.01	4.03	4.05
210	4.05	4.08	**4.10**	4.12	4.15	4.17	4.19	4.21	4.24	4.26	4.28

Type K Thermocouples
Temperature in Degrees F   Reference Junction at 32 F
EMF in Millivolts

°F	Millivolts	°F	Millivolts	°F	Millivolts	°F	Millivolts
210	4.05	360	7.42	510	10.79	660	14.25
220	4.28	370	7.64	520	11.02	670	14.48
230	4.51	380	7.87	530	11.25	680	14.71
240	4.74	390	8.09	540	11.78	690	14.95
250	4.97	400	8.31	550	11.71	700	15.18
260	5.20	410	8.54	560	11.94	710	15.41
270	5.42	420	8.76	570	12.17	720	15.65
280	5.65	430	8.98	580	12.40	730	15.88
290	5.87	440	9.21	590	12.63	740	16.12
300	6.09	450	9.43	600	12.86	750	16.35
310	6.31	460	9.66	610	13.09	760	16.59
320	6.53	470	9.88	620	13.32	770	16.82
330	6.76	480	10.11	630	13.55	780	17.06
340	6.98	490	10.34	640	13.78	790	17.29
350	7.20	500	10.57	650	14.02	800	17.53

Type K Thermocouples

Temperature in Degrees F   Reference Junction at 32 F
EMF in Millivolts

°F	Millivolts	°F	Millivolts	°F	Millivolts	°F	Millivolts
800	17.53	1210	27.22	1710	38.65	2120	47.41
810	17.76	1215	27.34	1720	38.87	2130	47.62
820	18.00	1220	27.45	1730	39.09	2140	47.82
830	18.23	1230	27.69	1740	39.31	2150	48.03
840	18.47	1240	27.92	1750	39.53	2160	48.23
850	18.70	1250	28.15	1760	39.75	2170	48.44
860	18.94	1260	28.39	1770	39.96	2180	48.64
870	19.18	1270	28.62	1780	40.18	2190	48.85
880	19.41	1280	28.86	1790	40.40	2200	49.05
890	19.65	1290	29.09	1800	40.62	2210	49.25
900	19.89	1300	29.32	1810	40.84	2220	49.45
910	20.13	1310	29.56	1820	41.50	2230	49.65
920	20.36	1320	29.79	1830	41.27	2240	49.86
930	20.60	1330	30.02	1840	41.49	2250	50.06
940	20.84	1340	30.25	1850	41.70	2260	50.26
950	21.07	1350	30.49	1860	41.92	2270	50.46
960	21.31	1360	30.72	1870	42.14	2280	50.65
970	21.54	1370	30.95	1880	42.35	2290	50.58
980	21.78	1380	31.18	1890	42.57	2300	51.05
990	22.02	1390	31.42	1900	42.78	2310	51.25

1000	22.26	1400	31.65	1910	42.99	2320	51.45
1010	22.49	1510	34.16	1920	43.12	2330	51.64
1020	22.73	1520	34.39	1930	43.42	2340	51.84
1030	22.97	1530	34.62	1940	43.63	2350	52.03
1040	23.20	1540	34.84	1950	43.85	2360	52.23
1050	23.44	1550	35.07	1960	44.06	2370	52.42
1060	23.68	1560	35.29	1970	44.27	2380	52.62
1070	23.91	1570	35.52	1980	44.49	2390	52.81
1080	24.15	1580	35.75	1990	44.70	2400	53.01
1090	24.39	1590	35.97	2000	44.91	2410	53.20
1100	24.63	1600	36.19	2010	45.12	2420	53.39
1110	24.86	1610	36.42	2020	45.33	2430	53.59
1120	25.10	1620	36.64	2030	45.54	2440	53.78
1130	25.34	1630	36.87	2040	45.75	2450	53.97
1140	25.57	1640	37.09	2050	45.96	2460	54.16
1150	25.81	1650	37.31	2060	46.17	2470	54.35
1160	26.05	1660	37.54	2070	46.38	2480	54.54
1170	26.28	1670	37.76	2080	46.58	2490	54.73
1180	26.52	1680	37.98	2090	46.79	2500	54.92
1190	26.75	1690	38.20	2100	47.00		
1200	26.98	1700	38.43	2110	47.21		

SUPPLIERS:

American Foundrymen's Society
505 State Street
Des Plaines, IL 60016-8399

Foundry books, from basic to advanced
800 537 4237 www.afsinc.com
moderncastaing.com

Hickman Williams
US 800 862-1890
Canada 800 265 6415
Mexico 011 52 8 363-4041

St. Louis Coke & Foundry Supply
314 727-7500

Mifco
700 Griggs St.
Danville, IL 61834
217 446-0941
Thermocouples, letters, ladle wash, tools

Joy-Mark Inc.
2121 E Norse Ave.
Cudahy, Wis. 53110
414 769-8155
Ceramic reinforced ladles.
Part# 407   10 x 8.5 x 11.375

This ladle is very light. I use it for all my iron and aluminum work. It is highly recommended.

Insulating riser sleeves, riser toppings

Lindsay Publications
P.O. Box 538
Bradley, IL 60915

815 935 - 5353
lindsaybks.com
(Foundry books , Technical books)

Home Shop Machinist Magazine
2779 Aero Park Drive
Traverse City, MI 49686

Magazine for beginning and intermediate machinists

MSC Industrial Supply Co.
75 Maxess Road
Melville, NY 11747-9415
(800) 645-7270

Morganite Crucible
P. O. Box 338
North Haven, CT 06473
(203) 697-0808
morganitecrucibleinc.com

Budget Casting Supply
60 East 40th Ave.
San Mateo CA 94403
650 345-3891
budgetcastingsupply.com

Piedmont Foundry Supplies
3191 Rogers Lane
Cloverdale, VA 24077
Phone 992-3911
Columbus,GA Phone 324-5938

Porter Warner Foundry Supplies:

Alabama   205 251-8223
Arizona   602 244-9166

Tennessee 423 266-4735

California   213 722-1335
                   501 633 0285
                   909 822 5591

TX   817 633-1767
       713 222 6210

South Carolina 803 789-3444

Uni. West Foundry Supplies:

Denver, CO 303 388-0922   Washington 206 767-9880

Oregon 503 226-4836

Rice Industries
Minnesota 651 784-1881

Metal Ingot – all types

Belmont Metals
belmontmetals.com
718 342 4900

Atlas Metal Sales
1401 Umatilla St.
Denver, CO 80204 - 2432
303 623 0143

Ferro silicon, Nickel, Copper Phosphorus:
ASi International 1440 E. 39th St., Cleveland, Oh 44114 (216) 391-9900

Seacoal, Blacking, Coke:
Empire Coke 1927 1st Ave North #900, Birmingham, AL 35203 (205) 323-2400

Chaplets: Smith & Richardson, P.O. Box 589, Geneva, IL 60134 (630) -232-2581

Bentonite:
American Colloid Co.
1500 W. Sure Drive
Arlington Hts. IL 60004 (847) 392-4600

Wyo-Ben Inc.
P.O. Box 1979
Billings, MT 59103 (406) 652 6351

Chromite, Olivine, and Zircon Sands:
American Minerals, 901 E. 8th Ave, King of Prussia, PA 19406, (610) 337-8030

Sand & Core Binders:

Ashland Casting Solutions
P.O. Box 2219
Columbus, OH 43216
(614) 790-3346

Southern Foundry Resins
16 Belcher Dr.
Pelham, AL 35124
(205) 664-2255

Petro-Bond:
Bentonite Performance Minerals
250 N Sunny Slope Rd. Suite 204
Brook Field, WI 53005
(262) 786-5886

Filters – Refractory Cloth:
Ametek
900 Green Bank Rd.
Wilmington, DE 19808
800 - 441-7777
www.ametek.com/haveg

Precision Parallels, inexpensive Wood and Metal working Tools:

Harbor Freight Tools
3491 Mission Oaks Blvd.
Camarillo, CA 93011-6010
1 800 423- 2567

Harborfreight.com

Enco Tools
400 Nevada Pacific Hwy.
Fernly, NV 89408
1 800-USE ENCO (873-3626)

www.use-enco.com

INDEX:

Aircraft, Cylinder Casting, 160-161
Bentone, 153
Bentonite,
  Western & Southern, 145
Blower
  Furnace, shutter, 80
Boron Nitride, Coating, use of, 82
Brass Shakes, Causes of, 8
Cam, Furnace lid, Drawing, 90, 92
Cereal, 144, 170
Cement, as binder, 150-152
Charging, Hopper, 94
Chill, 49,53,57
Chromite, 126, 141
Core,
  Arbor, 167
  Baking, 173
  Binders,169
  Binder, order of addition, 171-172
  Box, 23, 25
    Defined 163-164, 166
  Buoyancy
    Calculating, 185
    Defined, 185
  Coatings, 193-195
  Cylinder Casting, Sand mix, 190
  Drier, 164,165
  Dry Sand, 160
  Engine, 162, 164-165, 180-184
  Green Sand, 160
  Ingredients
    Properties of and
    Ratio of, 189,190
    Weight of,188
  Metal Penetration of, 193
  Mixer, 172
  Molding, for vibrator, 29
  Piston, 162
  Prints, 24, 161-162, 186-187
  Properties of, 168
  Rollover Machine, 168
  Venting, 165-166
  Wash, 193
Chaplets, 186-188
Charcoal
  Facing, 152
  Furnace, 75-83
  Splitting, 82-83
Cheeks, Flask, 10, 13
Clay, 145-146
  Bonding, 146
  Saturation147

Combustion,
  Cyclone, 86
Cope, 9
Corn Flour, 170
Crankcase, Corebox, 164
Crucibles, 61-65
  Steel,
    Coated, 65, 82
    Making, 80-82
Cupola, 72-74
  Tapping,
    Photo- front cover
Density
  Molten Metals, 69
Dextrin, 144
Draft, Pattern, 23
Drag, 9
Dry Strength, of Sand, 127
Emissivity, of Materials, 124,125
Exhaust port, furnace, 87
Finger Joints, Wood, Cutting, 41-45
Fireclay, 145
Flask, 9-14
  Aluminum, Making, 54-57
  Hardware, 45-52, 59-60
  Snap, 10, 12
  Pop Off, 10
  Wood, Making, 40-46
Flowability, of Sand, 127
Follow Board, 16
Ford, Engine Core, 182
Form, Refractory, 77,79
Foundry Rule, Fundamental, 126
Fuel Oil, Addition of, 144
Furan Binders, 158-159
Furnace
  Crucible Type, 71
  Direct Fired, 70
  Gas, building, 84
  Indirect Fired, 70
  Lid, Making, 88-87, 93
  Lighting, 80, 97
  Reverberatory. 70
Grain Fineness Number, 133-134
Graphite, 144
Green Sand, 9
Green Strength, 127
Grooves, Flask, Sand Retaining, 10
Heater, Ladle 165
Heat Loss, from ladle, 66, 67
Hopper, Charging, 94
Insert, in casting, Steel, 28
Jacket, Mold, 14
Joy-Mark, 67

Ladle, 61
  Capacity, calculating, 68-69
  Explosion, 65
  Heater, 65
  Heat Loss From, 66-67
  Liner, 67
Lead, Cores Held by, 181-182
Letters, pattern, 54,56
Lighting, Furnace, 80, 97
Lining, Furnace, Firing, 80
Linseed Oil, 171
Mack, Diesel Engine Cores, 184
Metals, Properties of, Table, 199
Molasses, 144, 169
Mold,
  Making, 29
  Pouring, Photo of, 35
  Ramming, 31
  Striking, 32
Molder's Tools, 18
Molding Board, 16
Morganite,
  Address, 197
  Crucible, 62
Mulling Time, 149
Mullers, 148-150
Oil, Bonded Sand, 153-154
Olivine, 126, 134-141
Orifice
  Air, Location of, 36
  Gas, 85
Packard V-12 Engine Cores 183-184
Parting Dust, 19
Permeability, of Sand, 127, 132
Petro Bond, 154-157
Plinth, Furnace, Making, 95
Plumbago, 144
Pop Off Flask, 10
Pouring
  Basin, 9, 13
  Cup, 9, 36
Propane, Tanks, 96
Pyrometer, Optical 121-124
Rammer, 57-58
Rapping Bar, 17-18
  Use of, 34
Regulators, Gas, 95
Riddle, Sand, 17, 30
Sand
  Compound Grains, 128-129
  Compounded Mixture, 136-137
  Distribution, 134-135
  Grains, 128
    Shape 129

  Size, 129, 131
  Surface Area, 130-132
Mixes, 138-139
Naturally Bonded, 135, 136
Retaining Grooves, Flask, 10
Synthetic, 136-137
Thermal Expansion, 137
Wedge Test, 136
Sea Coal, 142
Sheet Metal, Forming, 98
Shrink Bob, 54
Silica, 126, 137
  Disease Caused by, 7
  Flour, 143
Silicon Carbide,
  Crucibles, 61, 64
  Sand, 126, 141
Silicosis, 7
Skimmer, Dross, 19-20
Snap Flask, 10, 12
Sodium Silicate, use in Cores, 160
Southern Bentonite, 145
Sprue, 9
  Cutting, 33
Steel, Crucible, Making, 80-82
Suppliers, List of 197,198
Tanks, Propane, 96
Temperature
  Color Scale, 113
  Measurement, 113
Thermocouple, 114-119
  Correcting for
    Ambient Temperature, 116-117
  Making, 117, 121
  Reference Junction, 115
  Tables, K type, 200-203
Tilting Furnace, 72
Transition Piece, Sheet Metal, 98-112
Upset, Flask, 11
V-12 Engine Cores, 183-184
Valve, Ball, Furnace, 84
Vaporization Capacity
  of Tanks, 96
Vent Rod, 18-19
Venting, 35
Vibrator, Match Plate, 21-22, 39
Wash, Ladle, 82
Well, 9
Western Bentonite, 145
Wet Strength, of Sand, 127
Wheat Flour, 170, 189
Wood Flour, 142, 144
Zinc Oxide, Inhaled, 8
Zircon, 126, 141

A Word to Beginners:

You do not have to spend a large sum of money to set up a small foundry and start casting metal. My first aluminum was melted in a welded steel crucible made by a local tech school. A charcoal-fired furnace was built with sand dug from the ground and mixed with fireclay left over from a neighbor's fireplace. The molten aluminum was poured into a pile of sand and clay (I had no flask) that had been tamped to a flat surface and had several holes poked into it using a 1 ½-inch diameter wooden dowel. After a few heats, I had enough ingots to pour a mold. My first pattern was carved from a dry oak tree limb that had fallen in my yard. Clearly, I had little invested in my first castings.

The storage space required for a small foundry is minimal. You can mix up a few hundred pounds of sand and store it outside in a plastic garbage can with wheels. The furnace, tools and a few flasks fit under a workbench.

Sometimes things will not go as planned. Do not be discouraged with a few failures. You are learning something new and a few setbacks are to be expected. Study your castings and process carefully, soon you will be able to produce good castings without fail.

A good beginner's casting project that you can build with a few hand tools is the "Metal Lathe" in the series "Build your Own Metal Working Shop from Scrap" by Dave Gingery. ISBN 1-878087-01-1.

Good luck with your casting projects. "The Sand Casting Manual" is continued in volume 2.

SMALL FOUNDRY FURNACES VOLUME 2: BUILD A ROTATING FURNACE FOR IRON MELTING

Volume 2 contains plans for a stand alone batch melter or "cupola receiver" that accumulates taps for larger pours. It continues with detailed information for building higher temperature burners and furnaces. Design information is expanded to include natural gas, high pressure propane, and oils. Atomizing burners and blending of waste oils is also described.

ISBN 13: 978-0-9702203-7-0   128 Pages $19.95

SAND MACHINERY VOLUME 1: BUILD A MULLER

Plans for a high quality sand muller to take the labor out of mixing your foundry sands. Good for sand-clay mixtures or Petrobond. Design information is included so that you may scale it up or down to match your foundry needs.

ISBN 13: 978-0-9702203-8-7   128 pages $19.95

MAKING PISTONS FOR RESTORATION ENGINES

Bring even the most "impossible" old engines back to life for little cost! You are no longer limited by the price and availability of replacement pistons and rings when you can make your own. Use inexpensive modern piston rings on your antique equipment! Design and make pistons for new or old engines. Learn to make all the tools and jigs needed to quickly produce top quality replacements in your own back yard. Heavily illustrated. A "must have" for antique equipment restorers!

ISBN 09702203-4-0 ,   ISBN 13: 978-0-9702203-4-9   64 Pages $11.95

GENERATORS AND INVERTERS: BUILDING SMALL COMBINED HEAT AND POWER PLANTS FOR REMOTE LOCATIONS AND EMERGENCY SITUATIONS

The most complete small generator reference and how-to book available! Everything that you need to know about building or buying small power plants. If you have any questions about generating your own electricity, you need this book! Not only does Steve answer your questions, he converts an auto engine into a power plant that runs his house and heats his pool.

ISBN 13: 978-0-9702203-5-6   352 pages $34.95

Color Photos at: StephenChastain.Com

The Small Foundry Series:

Metal Casting: A Sand Casting Manual Vol. 1 & 2

Learn how to cast metal in sand molds using simple techniques and readily available materials. Steve Chastain, a Mechanical and Materials Engineer shows the beginner how to make a sand mold and then how to hone your skills to produce high quality castings. Written in non-technical terms, the sand casting manuals begin by melting aluminum cans over a charcoal fire and end by casting a cylinder head. All for little cost in your own back yard! Includes the basic metallurgy of aluminum, iron and copper-based alloys. Sold in over 45 countries, Chastain's popular "Small Foundry Series" is good for both the beginner and experienced metal caster.

Volume I   ISBN 0-9702203-2-4 ,   ISBN 13: 978-0-9702203-2-5

208 Pages $19.95

Volume II   ISBN 0-9702203-3-4,   ISBN 13: 978-0-9702203-3-2

192 Pages $19.95

Iron Melting Cupola Furnaces

Complete plans and operating instructions for a 10" diameter cupola that will melt 330 pounds of iron per hour when powered by a shop vac! Also included are plans for a high-pressure blower that will increase the output of the small furnace to 660 pounds per hour. All can be built for little cost, mostly from scrap.

ISBN 0-9702203-0-8 , ISBN 13: 978-0-9702203-0-1   128 pages $19.95

Small Foundry Furnaces Volume 1:
Build a Tilting Furnace

Complete plans and operating instructions for a tilting furnace that easily melts 100 pounds of aluminum per hour. Melt with propane, diesel, or used motor oil! Using the tilting mechanism, you will never have to handle a hot crucible again. Start, stop, and hold the furnace at any angle for precise pours. Furnace may be modified to melt other metals.

ISBN 09702203-1-6, ISBN 13: 978-0-9702203-0-1   192 Pages $19.95

Continued on page 207   StephenChastain.Com